BIOASSAY METHODS IN NATURAL PRODUCT RESEARCH AND DRUG DEVELOPMENT

Proceedings of the Phytochemical Society of Europe

Volume 43

Bioassay Methods in Natural Product Research and Drug Development

Edited by

Lars Bohlin

and

Jan G. Bruhn

Division of Pharmacognosy and
Centre for Applied Pharmacognosy,
Uppsala University,
Uppsala, Sweden

SPRINGER - SCIENCE+BUSINESS MEDIA, B.V.

A C.I.P. Catalogue record for this book is available from the Library of Congress.

ISBN 978-0-7923-5480-2 ISBN 978-94-011-4810-8 (eBook)
DOI 10.1007/978-94-011-4810-8

Printed on acid-free paper

Preface

This book contains the proceedings of the international symposium on 'Bioassay Methods in Natural Product Research and Drug Development', held at the Biomedical Centre of Uppsala University, Sweden, from 24-27 August 1997. This symposium was organized by the Phytochemical Society of Europe in collaboration with the Swedish Academy of Pharmaceutical Sciences. On behalf of the scientific and organizing comittee we want to thank the Society and the Academy for their support and we are also very grateful for the generous sponsorship provided by the research councils and companies listed after this preface.

The symposium attracted 200 delegates from more than 30 different countries, a good testimony to the world-wide interest in drug discovery from natural sources. We have had very many positive comments from colleagues participating in the symposium, about the high quality of lectures and discussions, but also in praise of the social atmosphere of the meeting, which has lead to many new friendships and collaborative efforts.

The purpose of the meeting was to focus on, and discuss the further development of strategies and methods for selection of organisms, extraction procedures, bioassay-directed isolation and characterization of natural products. The chances of isolating unique molecules with new modes of action were discussed, as well as how to avoid the replication of known chemical structures with already established biological activity. The programme reflected both academic and industrial research.

We were fortunate to attract the foremost leaders in the field, and we believe that the reader of this text will recognize the high standard of the meeting in every chapter. The organization of the book follows that of the meeting, discussing first biological and chemical diversity, and then presenting biological, pharmacological and chemical evaluation of natural products. This is followed by chapters on natural products with chemotherapeutic potential and immunomodulating activity, and then examples of applied industrial research in the drug discovery process. The final contributions outline strategies for future research.

The scientific and organizing comittee consisted of Stig Agurell, Peter Brodelius, Anders Cronlund, Torbjörn Norin, Kirsi-Marja Oksman-Caldentey, Wenche Rolfsen, Robert Verpoorte and the undersigned. We thank all the speakers for their valuable contributions to the meeting and to this volume, which will certainly stimulate further research.

Uppsala, July 1998

Lars Bohlin and Jan G Bruhn
Division of Pharmacognosy and Centre for Applied Pharmacognosy,
Faculty of Pharmacy, UppsalaUniversity,
Box 579, SE-751 23 UPPSALA, SWEDEN

Sponsorship

We are greatly indebted to the following companies and institutions for their generous support during the international symposium on Bioassay Methods in Natural Product Research and Drug Development" held at the Biomedical Centre of Uppsala University, Sweden, 24-27 August 1997.

Astra Läkemedel AB, Södertälje, Sweden

BioPhausia AB, Uppsala, Sweden

Conpharm AB, Uppsala, Sweden

Pharmacia Biotech AB, Uppsala, Sweden

Swedish Council for Forestry and Agricultural Research, Stockholm, Sweden

Swedish Natural Science Research Council, Stockholm, Sweden

Swedish Research Council for Engineering Sciences, Stockholm, Sweden

CONTENTS

Contributors

R. Bauer: Institut für Pharmazeutische Biologie der Universität Düsseldorf, Universitätsstr 1, D-40225 Düsseldorf, Germany

A. J. J. van den Berg: Utrecht University Department of Pharmacy, Section of Medicinal Chemistry, Pharmacognosy Unit, F.A.F.C. Went Building, Sorbonnelaan 16, NL-3584 CA Utrecht, The Netherlands

C. J. Beukelman: Utrecht University Department of Pharmacy, Section of Medicinal Chemistry, Pharmacognosy Unit, F.A.F.C. Went Building, Sorbonnelaan 16, NL-3584 CA Utrecht, The Netherlands

Jamie Boswell: Laboratory of Drug Discovery Research and Development, Developmental Therapeutics Program, Division of Cancer Treatment and Diagnosis, National Cancer Institute, Frederick Cancer Research and Development Center, P.O. Box B. Bldg. 1052, Room 121, Frederick, MD 21702-1201

Michael R. Boyd: Laboratory of Drug Discovery Research and Development, Developmental Therapeutics Program, Division of Cancer Treatment and Diagnosis, National Cancer Institute, Frederick Cancer Research and Development Center, P.O. Box B. Bldg. 1052, Room 121, Frederick, MD 21702-1201

Antony D. Buss: Bioprocessing Research Unit, Glaxo Wellcome Research and Development, Medicines Research Centre, Gunnels Wood Road, Stevenage, Hertfordshire SG1 2NY, UK

John H. Cardellina II: Laboratory of Drug Discovery Research and Development, Developmental Therapeutics Program, Division of Cancer Treatment and Diagnosis, National Cancer Institute, Frederick Cancer Research and Development Center, P.O. Box B. Bldg. 1052, Room 121, Frederick, MD 21702-1201

S. L. Croft: London School of Hygiene and Tropical Medicine, Department of Infectious and Tropical Diseases, Keppel Street, London WC1E 7HT, United Kingdom

Michael Currens: Laboratory of Drug Discovery Research and Development, Developmental Therapeutics Program, Division of Cancer Treatment and Diagnosis, National Cancer Institute, Frederick Cancer Research and Development Center, P.O. Box B. Bldg. 1052, Room 121, Frederick, MD 21702-1201

H. van Dijk: Utrecht University Department of Pharmacy, Section of Medicinal Chemistry, Pharmacognosy Unit, F.A.F.C. Went Building, Sorbonnelaan 16, NL-3584 CA Utrecht, The Netherlands

T. Elmquist: Scientific Research Director, Swedish Biodiversity Centre, Box 7007, SE-750 07, Sweden

Richard W. Fuller: Laboratory of Drug Discovery Research and Development, Developmental Therapeutics Program, Division of Cancer Treatment and Diagnosis, National Cancer Institute, Frederick Cancer Research and Development Center, P.O. Box B. Bldg. 1052, Room 121, Frederick, MD 21702-1201

William R. Gamble: Laboratory of Drug Discovery Research and Development, Developmental Therapeutics Program, Division of Cancer Treatment and Diagnosis, National Cancer Institute, Frederick Cancer Research and Development Center, P.O. Box B. Bldg. 1052, Room 121, Frederick, MD 21702-1201

S. B. A. Halkes: Utrecht University Department of Pharmacy, Section of Medicinal Chemistry, Pharmacognosy Unit, F.A.F.C. Went Building, Sorbonnelaan 16, NL-3584 CA Utrecht, The Netherlands

K. Hostettmann: Institut de Pharmacognosie et Phytochimie, Université de Lausanne, B.E.P., CH-1015 Lausanne, Switzerland

Peter J. Hylands: Oxford Natural Products PLC, No. 1 St Giles, Oxford OX1 3JS, United Kingdom

B. H. Kroes: Utrecht University Department of Pharmacy, Section of Medicinal Chemistry, Pharmacognosy Unit, F.A.F.C. Went Building, Sorbonnelaan 16 NL-3584 CA Utrecht, The Netherlands

R. P. Labadie: Utrecht University Department of Pharmacy, Section of Medicinal Chemistry, Pharmacognosy Unit, F.A.F.C. Went Building, Sorbonnelaan 16 NL-3584 CA Utrecht, The Netherlands

A. Marston: Institut de Pharmacognosie et Phytochimie, Université de Lausanne, B.E.P., CH-1015 Lausanne, Switzerland

James B. McAlpine: Phytera, Inc., 377 Plantation Street, Worcester, MA 01606, USA

J. H. van Meer: Utrecht University Department of Pharmacy, Section of Medicinal Chemistry, Pharmacognosy Unit, F.A.F.C. Went Building, Sorbonnelaan 16, NL-3584 CA Utrecht, The Netherlands

Murray H. G. Munro: Laboratory of Drug Discovery Research and Development, Development Therapeutics Program, Division of Cancer Treatment and Diagnosis, National Cancer Institute, Frederick Cancer Research and Development Center, P.O. Box B. Bldg. 1052, Room 121, Frederick, MD 21702-1201

Christopher Pazoles: Phytera, Inc., 377 Plantation Street, Worcester, MA 01606, USA

R. J. Quinn: Queensland Pharmaceutical Research Institute, Griffith University, Brisbane, 4111, Australia

W. N .A. Rolfsen Sandsborg: Division of Pharmacognosy, Department of Pharmacy, Biomedical Center, Uppsala University, Box 579, SE-751 23 Uppsala, Sweden and Phase I Services, Quintiles AB, Islandsgatan 2, SE-753 18 Uppsala, Sweden

H. F. Smit: Utrecht University Department of Pharmacy, Section of Medicinal Chemistry, Pharmacognosy Unit, F.A.F.C. Went Building, Sorbonnelaan 16, NL-3584 CA Utrecht, The Netherlands

Angela Stafford: Phytera Ltd., Regent Court, Regent Street, Sheffield, S14DA, UK

Tineke L. Tinbergen de Boer: Utrecht University Department of Pharmacy, Section of Medicinal Chemistry, Pharmacognosy Unit, F.A.F.C. Went Building, Sorbonnelaan 16, NL-3584 CA Utrecht, The Netherlands

Martin Th. M. Tulp: Solway Duphar B.V., Department of Pharmacology, P.O. Box 900, 1380 DA Weesp, The Netherlands

Linda C. Quarles van Ufford: Utrecht University Department of Pharmacy, Section of Medicinal Chemistry, Pharmacognosy Unit, F.A.F.C. Went Building, Sorbonnelaan 16, NL-3584 CA Utrecht, The Netherlands

R. Verpoorte: Division of Pharmacognosy, Leiden/Amsterdam Center for Drug Research, Leiden University, Leiden, The Netherlands, Email: verpoorte@LACDR.LeidenUniv.NL

A. J. Vlietinck: Department of Pharmaceutical Sciences, University of Antwerp (UIA), Universitetsplein 1, B-2610 Antwerp, Belgium

H. Wagner: Institute of Pharmaceutical Biology, University of Munich, Karlstrasse 29, D-80333, Munich, Germany

C. R. Weiss: London School of Hygiene and Tropical Medicine, Department of Infectious and Tropical Diseases, Keppel Street, London WC1E 7HT, United Kingdom

Chandra Westergaard: Laboratory of Drug Discovery Research and Development, Developmental Therapeutics Program, Division of Cancer Treatment and Diagnosis, National Cancer Institute, Frederick Cancer Research and Development Center, P.O. Box B. Bldg. 1052, Room 121, Frederick, MD 21702-1201

E. van den Worm: Utrecht University Department of Pharmacy, Section of Medicinal Chemistry, Pharmacognosy Unit, F.A.F.C. Went Building, Sorbonnelaan 16, NL-3584 CA Utrecht, The Netherlands

PLANT BIODIVERSITY -
Evolutionary and Ecological Perspectives

T. ELMQVIST
Scientific Research Director
Swedish Biodiversity Centre, Box 7007, S-750 07 Uppsala, Sweden

Abstract

Biodiversity is the totality of variation in life-forms, across all levels of biological organization, from genes within species to the array of species themselves and finally, at the highest level, the living communities of ecosystems. The Rio Convention on Biological Diversity deals predominately with three crucial issues: ***the identification, conservation*** and ***sustainable use of biodiversity.*** These issues are currently the focus of intensive research around the world e.g. on the characterization of genetic diversity and processes that generate and maintain biodiversity, classification of ecosystems/habitats, the role of diversity for ecosystem function, the role of man in maintenance of biodiversity, and the links between domesticated and wild biodiversity.

The current very rapid biotechnology development has resulted in that the world´s genetic and species diversity increasingly are viewed as industrial assets of great potential economic value. New search strategies for identifying valuable areas or taxa for biodiversity prospecting are rapidly under development, using a variety of phylogenetic, biogeographic, genetic, and ecological data.

1. Introduction. What is biodiversity and how do we measure it?

The recent decade has witnessed an exploding usage of the term biodiversity and so far uncompleted attempts to find a rigorous definition (Harper and Hawksworth 1995, Gaston 1996). Based on the definition used in the Convention on Biological Diversity (UNEP 1992), Harper and Hawksworth (1995) suggested that biodiversity represents "variability among living organisms, including genetic, organismal and ecological variation". It is not clear from these definitions whether processes which are of importance for the evolution or maintenance of entities like genes, individuals or species are explicitly included or not. In recent use of the term, however, such evolutionary or ecological processes are often viewed as an integral part (e.g. Gaston 1996).

The choice of definitions will ultimately be linked to unambiguous methods of quantification of diversity at different levels. Finding a single method which embraces all facets and complexity will prove difficult, if not impossible. The most widely used method are measures of species richness, which have the advantage of being relatively easy to perform and interpret, but which encompass only a fraction of the whole biodiversity. Just counting the number of species in an area will give a poor representation of the overall biodiversity and may sometimes be misleading on which areas will have the highest conservation priority. All species do not have the same weight. For example, a loss of a species from a very species - rich genus would detract less from overall biodiversity than the loss of a species from a single species genus. In some instances, higher taxa (phyla, orders and families) might provide more appropriate measures of biodiversity. Estimates of taxonomic distances by analyzing how many kingdoms are present at a site, how many phyla are represented per kingdom and how many orders per phylum will give a different but perhaps much more informative measure of biodiversity (Harper and Hawksworth 1995). Furthermore, if the goal is to preserve the richest diversity of pharmacological compounds, family or endemic richness may be more appropriate than species richness. Other measures would for example include variation at the DNA and RNA level, measures that have the advantage of providing a commodity by which all organisms can be compared.

2. Patterns of Plant Diversity

At the genetic level, plants tend to maintain higher levels of allozyme variation within their populations than either invertebrates or vertebrates. Long-lived perennials tend to be more genetically diverse than short-lived perennials and wide-spread species are more diverse than narrowly distributed species (e.g. Hamrick and Godt 1990, WCMC 1992, Frankel et al. 1995). When analysing plant species diversity it is evident that the majority of species occur in the tropics with more than 66% of all the species and several families (e.g. Annonaceae, Moraceae, Dipterocarpaceae, Ebenaceae, Meliaceae) restricted to the tropics. Spatial patterns of plant diversity may be described by using indices based on different scales. Vegetation can be rich in species either because individual plant communities are rich in total number of species (high α-diversity), or because the spatial variability of the physical habitat influences species composition so that a mosaic of communities exists whose total species composition is high (high β- and γ - diversity).

The richest continent for plants, and still the least explored botanically, is South America, home to perhaps as much as one-third of the world's higher plants (with high α- β- and γ-diversity) (Table 1). Another area with very high plant species richness is the Cape region of Southern Africa, where species densities are very high (β- and γ - diversity being high) (Table 2). Within the tropics the most diverse species assemblages occur in lowland areas with high and evenly distributed rainfall (Gentry 1992). In these areas soil fertility appears to have a minor direct effect on species richness. Ecological factors important for maintaining high plant species diversities include abiotic factors such as fires and other disturbances, and biotic factors such as grazing by animals and pathogens. High predictability of environmental variation at all time scales seems, however, to be the universal requirement of high species richness in plant communities.

Table 1. Vascular plant diversity (from WCMC 1992)

Group	No.of species	Centre of diversity
Pteridophytes	12,000	Papua New Guinea (13%)
Gymnosperms	600	Oceania
Angiosperms	250,000[1]	South America (30%)

[1]Estimates vary between 250.000-350.000 species (Heywood 1995)

IUCN (The World Conservation Union) is currently working with identifying major Centres of Plant Biodiversity (CPDs) and three kinds of high-diversity sites are recognized: 1) botanically rich sites that can be recognized geographically, like Mt. Kinabalu in Borneo; 2) geographical regions with high species richness and/or endemism such as the Atlas Mountains and the Cape province; 3) vegetation types and floristic provinces that are exceptionally rich in plant species like the Amazonian basin and south-western Australia. The selection of sites is based on a set of criteria, which includes: a) areas known to contain a large number of endemic species; b) sites containing an important gene pool of plants of value to man or plants that are potentially useful; c) sites containing a significant proportion of species adapted to special edaphic conditions. So far, 241 Centres of Plant Diversity have been selected, most of them in the tropics (WCMC 1992).

Table 2. Regional species richness in a selected set of regions (WCMC 1992)

Area	No. of species	Size	Species/1000 km²
Tropical Africa	30,000	20,000	1,5
Brazil	40,000	8,456	4,7
Australia	25,000	7,716	3,2
Sweden	1,165	1,400	1,0
Cape region	8,500	90	94,5
Southern Africa	22,900	2,573	8,9

Another geographical ranking has been suggested by McNeely et al. (1990) who identified 12 megadiversity countries with high levels of species endemism: **Latin America:** Mexico, Colombia, Ecuador, Peru, Brazil; **Africa:** Zaire, Madagascar; Asia: China, India, Malaysia, Indonesia; **Oceania:** Australia. These countries alone harbour more than 70% of all the world´s biodiversity of vertebrates, higher plants and butterflies, and qualify as high priority targets for global conservation efforts.

However, our understanding of the factors governing species diversity and speciation processes is still limited. Speciation has for very long been viewed as a process of significance in situations of allopatry, i.e. a sub-population being geographically isolated from the rest of the species and over time becoming reproductively isolated. This view has, however, recently been challenged in a study from tropical Africa, indicating the possibility of sympatric speciation in ecotones (Smith et al. 1997).

2.1. ENDEMISM

Endemism, i.e. taxa restricted to a specified geographic region, represents a concept used for many different spatial scales and taxonomic levels. However, despite this shortcoming of relativity, quantification of endemism provides a powerful tool for identifying areas of high priority for conservation and in some cases, areas suitable for biodiversity prospecting. Attempts to view endemism from an evolutionary perspective were made by Engler already in the late 1800s. He defined *neoendemics* as clusters of closely related species and subspecies that have evolved relatively recently, and *paleoendemics* as phylogenetically high-ranking taxa, usually sections, subgenera and genera that may be regarded as evolutionary relicts (Heywood 1995).

Endemic biota are not random assemblages from a phylogenetic point of view. The fact that some plant families are more represented among endemic taxa than the average plant family (e.g. Proteaceae and Ericaceae) and some less represented (Poaceae and Cyperaceae) might indicate that some taxonomic groups are pre-adapted to cope with small population sizes and have abilities to persist in the face of habitat fragmentation (Lawton 1993).

Endemic species are non-randomly distributed around the world. Areas of endemism tend to be clustered, sometimes congruently across a wide range of taxa. This may be particularly marked on mountain tops and on oceanic islands (Heywood 1995) reflecting a high degree of isolation from source populations. In some cases congruence may be related to mutualism, e.g. between speciose groups of plants and specialized pollinators, as suggested for patterns of congruent endemism in Madagascar (e.g. Nilsson 1988). Among plants, high endemism is also related to substrates that are nutritionally imbalanced (e.g. serpentine, limestone). On such substrates, endemism is often predicted to be high also among plant symbionts. From an ecological perspective, endemic species may belong to any of four different types depending on the habitat specificity (broad or restricted) and local population size (somewhere large or everywhere small) (Rabinowitz et al. 1986). However, the abiotic and biotic correlates with endemism are still not very well understood.

The 18 most species rich areas in the world, commonly referred to as hot spots (Myers 1989) support nearly 50,000 endemic plant species - about 20% of the world's total flora - but comprise only 0.5% of the earth's surface. Fourteen areas are within tropical continental forests. Some examples of areas with very high endemism include the Cape region in South Africa and some oceanic islands in the South Pacific like New Caledonia.

2.1.1 The Cape Flora

There is no other place on earth which matches the stunning diversity of this relatively small nutrient poor area comprising about 90,000 km², but harbouring more than 8,600 species of which 5,800 are endemics. The Cape Flora contains more than 70% of the *Erica*-species (740) and 60% of *Gladiolus*- species (160) of the world (Cowling and Richardson 1995). The bulk of endemic plants are neoendemics belonging to Ericaceae, Proteaceae, Rutaceae, Rhamnaceae, Fabaceae, Polygalaceae and Rosaceae. In the dominating type of vegetation, the *Fynbos*, a shrubby dry vegetation with few tree species, the alpha diversity may reach values of 65 species per 1000 m², but this is not too exceptional in a global perspective. What is truly stunning is the extremely high beta-and gamma-diversity. Two closely located areas may share just a tiny fraction of the combined number of species. For example, the Kogelberg and The Cape Peninsula, two very similar areas separated by a narrow bay may share less than 50% of their species (Cowling and Richardson 1995). Dwarf succulent shrubs and geophytes are significantly over-represented among endemics.

No one has so far produced a convincing model to explain why the Cape Flora is so rich in species. There are however, certain biological traits in common for many of the plant groups showing high speciation, e.g. fire-sensitivity, low reproductive output, ant-dispersed seeds, small and weakly persisting seed banks. Together these factors indicate a high risk of fire causing drastic population declines and may result in high probabilities for genetic drift and subsequent speciation.

Today, alien invasive species constitute the largest threat to the 1326 Red Data Book plant taxa that are endemic in the Cape region and endemic species have often been found to be highly sensitive to the impact of exotic invasives, both among plant and animal.

2.1.2 Island Plant Endemism

Oceanic islands have distinct and often unique floras, frequently with high levels of endemism. Island endemics tend to be of two types, relict species which have had a broader distribution in the past and species which have evolved *in situ* as a result of isolation. The combination of low numbers of immigrants, infrequent immigration events due to isolation and small population sizes, can lead to high levels of genetic drift. Even after breeding populations of plants have been established on oceanic islands, the effects of reproductive isolation can be amplified by limited gene flow both between islands and between populations within a single island, due to the relative absence of pollinators and seed dispersers. All of these factors have resulted in a mosaic of island floras with high levels of endemism at the generic and specific level. New Caledonia has one of the most distinctive island floras in the world. Of 3322 vascular plants, 2551 (77%) are endemic, including five families and 110 genera. The island is particularly rich in conifers and 43 of the 44 species are endemic.

Island and mainland populations of angiosperms may differ in several aspects of their reproductive biology (Rick 1966, Carlqvist 1974). Comparisons between the breeding systems of island and mainland floras suggest that a greater proportion of island than of mainland plants is self-fertilizing (Baker 1955, Linsley 1966, Rick 1966, Baker and Hurd 1968). Plants with obligate outcrossing systems such as gametophytic self-incompatibility (Anderson and Stebbins 1984) or dioecism (Bawa 1982, Cox 1985) are less likely to establish breeding populations after long-distance dispersal, resulting in a sampling bias towards species capable of self-fertilization. Furthermore, the flowers of insect-pollinated plants on islands tend to have unspecialized open structures, that are smaller and less showy than mainland flowers (Carlqvist 1974). Within plant species, fruit set and seed set is often lower, or, show larger temporal variability on islands compared to mainland populations (Spears 1987). Wind pollination also appears to be more common on islands than in mainland populations (Whitehead 1969). The differences in reproductive characteristics may be a result of the impoverished pollinator communities usually found on islands (Baker 1955; Carlqvist 1966). For example, reduced diversity on islands has been found in insect communities (Simberloff & Wilson 1969) and birds (Diamond 1970) - two important pollinator groups - and in many other taxa; see reviews in MacArthur & Wilson (1967) and Carlqvist (1974). The few studies that have examined specific pollinator species on islands indicate that that they are scarcer or less predictable than on comparable mainland sites (Linsley 1966; Feinsinger et al. 1979; Linhart & Feinsinger 1980; Feinsinger et al.1982, Spears 1987, Elmqvist et al. 1992).

We know that 34% of all known threatened plant species in the world are island endemics and that the majority of these threatened plants are pollinated by a single or a few species of insects, birds or bats. Successful conservation, management, reintroductions and restoration, will thus require knowledge and protection of these plant-pollinator interactions.

3. Plant biodiversity, bioprospecting and search strategies

The current very rapid biotechnology development has resulted in that the world's genetic and species diversity increasingly are viewed as industrial assets of great potential economic value (Reid et al. 1993, Colwell 1997). Biotechnology can be used in a wide range of applications such as the search for novel genes and gene products to be used in bioremediation, biorestoration, breeding and using living organisms as factories delivering a variety of products (Myers 1997). Rapid development may, however, be hampered by the lack of cost-effective search strategies for novel genes and gene products. New search strategies for identifying valuable areas or taxa for biodiversity prospecting is rapidly under development, using a variety of phylogenetic, biogeographic, genetic and ecological data. This may complement other search strategies, such as the use of ethnobiological and ethnopharmacological information, which currently is increasingly used throughout the world (e.g. Reid et al. 1993, Balick and Cox 1996). Endemic taxa have often been used for identifying areas of high priority for conservation (Myers 1990). Endemic-rich areas often correspond with centres that are generally rich in species, particularly at larger spatial scales. Information about endemism may be useful also for biodiversity prospecting. Areas with high paleoendemism and with endemics growing under unusual edaphic conditions may be of particular interest in the search for novel genes and gene products.

4. Conclusion

The question central to all investigations of diversity, "Why are there so many species" was eloquently raised by G. E. Hutchinson already 1959 in a seminal paper entitled "Hommage to Santa Rosalia or Why are there so many kinds of animals?" Old questions like this now re-emerges in the process of understanding patterns of biodiversity together with new and intriguing scientific questions: In what way do taxonomic diversity and ecosystem processes interact? Are species that perform similar functions interchangeable and which biological entities are functionally most important? (e.g. Naeem et al. 1994, Tilman 1996, Tilman 1997). Does the diversity of species affect the sustainability of ecosystem function? (e.g. Daily 1997). How do we restore degraded ecosystems? (e.g. Jordan 1997). To solve these complex questions the **Science of Biodiversity** is today emerging as a scientific discipline with growing cooperation between usually far apart scientific disciplines such as physical sciences, social sciences and the humanities, geography and biology.

Acknowledgement

I thank Paul A. Cox, Torbjörn Ebenhard and Börge Pettersson for constructive criticism and fruitful discussions.

References

Anderson, G.J. and Stebbins, G.L (1984) Dioecy versus gametophytic self-incompatibility: a test, *American Naturalist* **24,** 423-428.

Baker H. G. (1955) Self-compatibility and establishment after long distance dispersal, *Evolution* **9,** 347-348.

Baker, H.G. and Hurd, P.D. Jr. (1968) Intrafloral ecology, *Annual Review of Entomology* **13,** 385-414

Balick M. J. and Cox P.A.(1996) *People, Plants and Culture,* Scientific American Library, New York.

Bawa, K. S. (1982) Outcrossing and the incidence of dioecism in island floras, *American Naturalist* **119,** 866-871,

Buchmann, S. and Nabhan, G. (1996) *The forgotten pollinators,* Island Press, Shearwater Books. Washington D.C.

Carlqvist, S. (1974). *Island Biology,* Columbia University Press, New York.

Colwell, R. R. (1997) Microbial Biodiversity and Biotechnology, in M. L. Reaka-Kudla, D. E. Wilson and E. O. Wilson (eds.), *Biodiversity II: understanding and protecting our biological resources,* National Academy of Science, USA. pp. 279-288.

Conner J.K. and Neumeier R. (1995) Effects on black mustard population size on the taxonomic composition of pollinators, *Oecologia* **104,** 218-224.

Cowling, R and Richardson, D. (1995) *Fynbos,* Fernwood Press.

Cox, P. A. (1985) Islands and dioecism: insights from the reproductive ecology of Pandanus tectorius in Polynesia, in J. White (ed.), *Studies on Plant Demography: A Festschrift for John L. Harper,* Academic Press, London, pp. 359-372.

Daily, G. (1997)Valuing and Safe-Guarding Earth's Life-Support System, in Daily, G. (ed.) *Nature's Services: Societal dependence on natural ecosystems,* Island Press, pp. 365-374.

Diamond, J. M. (1970) Ecological consequences of island colonization by Southwest pacific birds. II. The efect of species diversity on total population density. *Proceedings of the National Academy of Sciences,* **67,** 1715-1721.

Elmqvist, T., Cox, P., Rainey, W., Pierson, E. (1992) Restricted pollination on oceanic islands: Pollination of Ceiba pentandra by Flying Foxes in Samoa, *Biotropica,* **24,** 15-23.

Elmqvist, T, Rainey, W. E. Pierson, E. D. and Cox, P. A. (1994) The effects of two tropical cyclones on a lowland rain forest in Samoa, *Biotropica* **26**, 384-391.

Feinsinger, P., Linhart, Y. B. & Swarm, L. A. (1979) Aspects of the pollination biology of three *Erythrina* species on Trinidad and Tobago. *Annals of the Missouri Botanical Gardens;* **66,** 494-506.

Feinsinger, P., Wolfe, J. A. & Swarm, L. A. (1982) Island ecology: reduced hummingbird diversity and the pollination biology of plants, Trinidad and Tobago, West Indies, *Ecology,* **63,** 494-506.

Frankel, O.H. Brown, A. H. D. and Burdon, J. J. (1995) *The conservation of plant biodiversity,* Cambridge University Press.

Gaston, K.J. (1996) What is Biodiversity? in K. J. Gaston, *Biodiversity. A Biology of numbers and difference,* Blackwell Science, Oxford. pp. 1-12.

Gentry, A. H. 1992. Tropical forest biodiversity: distributional patterns and their conservational significance. *Oikos* **63,** 19-28.

Hamrick, J.L. and Godt, M.J. (1990) Allozyme diversity in plant species, in Brown, A.H. D., Clegg, M.T. Kahler, A.L. and Weir, B.S. *Plant population genetics, breeding and genetic resources.* Sinauer Ass. Inc. Sunderland, Mass.

Harper, J. L. and Hawksworth, D.L. (1995) Preface, in D.L. Hawksworth (ed.) *Biodiversity. Measurement and Estimation,* Chapman and Hall. London, pp.5-12.

Heywood V. H. (ed.) (1995) *Global Biodiversity Assessment,* UNEP, Cambridge University Press, pp. 174-185.

Hutchinson, G. E. (1959) Homage to Santa Rosalia or, Why are there so Many Kinds of Animals?, *American Naturalist* **93,** 145-159.

Jordan, W.R.III (1997) Ecological restoration and the Conservation of Biodiversity, in M. L. Reaka-Kudla, D. E. Wilson and E. O. Wilson (eds.), *Biodiversity II: understanding and protecting our biological resources,* National Academy of Science, USA. pp. 371-388.

Lawton, J. H. (1993) Range, population abundance and conservation. *Trends in Ecology and Evolution* **8,** 409-413.

Linhart, Y. & Feinsinger, P. (1980) Plant-hummingbird interactions: effect on island size and degree of specialization on pollination. *Journal of Ecology* **68,** 745-760.

Linsley, E. G. (1966) Pollinating insects of the Galapagos islands, in R. I. Bowman (ed.) *The Galapagos,* pp. 225-232. University of California Press, Berkley.

MacArthur, R. H. & Wilson, E. O. (1967) *The Theory of Island Biogeography,* Princeton University Press, Princeton, New Jersey.

McNeely, J. A., Miller, K.R., Reid, W.V., Mittermeier, R. A. and Werner, T.B. (1990) *Conserving the World's Biological Diversity,* IUCN, Gland, Switzerland.

Myers, N. (1990) The biodiversity challenge: expanded hot spots analysis. *The Environmentalist* **10,** 243-256.

Myers, N. (1997).Biodiversity's genetic library. pp. 255-274 in Daily, G. C. (ed.) *Natures services. Societal dependence on natural ecosystems,* Island Press, pp. 255-274.

Naeem, S., Thompson, L.J., Lawler, S.P., Lawton, J.H. and Woodfin, R.M. (1994) Decling biodiversity can alter the performance of ecosystems, *Nature* **368,** 734-737.

Nilsson, L. A. (1988) The evolution of flowers with deep corolla tubes. *Nature* **334,** 147-149.

Rabinowitz, D., Cairns, S. and Dillin, T. (1986) Seven forms of rarity and their frequency in the flora of the British Isles, in SoulÈ, M. (ed.), *Conservation biology. The science of scarcity and diversity,* Sinauer Ass., Sunderland, Mass, pp.182-204.

Reid, W. V., Laird, S. A., Meyer, C. A., G·mez, R., Sittenfeld, A., Janzen, D. H., Gollin, M. A., Juma, C. (1993), *Biodiversity prospecting.* World Resources Institute, USA, pp. 1-52

Rick, C. M. (1966) Some plant-animal relations on the Galapagos Islands, in R. I. Bowman (ed.), *The Galapagos,* University of California Press, Berkeley, pp. 214-224.

Simberloff, D.S. and Wilson, E.O. (1969) Experimental zoogeography of empty islands. *Ecology* **50,** 278-295.

Smith, T.B., Wayne, R.K., Girman, D. J. and Bruford, M. W. (1997) A role for ecotones in generating rainforest biodiversity, *Science* **276,** 1855-1857.

Spears, E.E. (1987) Island and mainland pollination, ecology of *Centrosema virginianum and Opuntia sticta, Journal of Ecology 75, 351-362.*

Tilman, D. (1996) Biodiversity: Population versus ecosystem stability. *Ecology* **77** 350-363.

Tilman, D. (1997) Biodiversity and Ecosystem Functioning, in Daily, G. (ed.) *Nature's Services: Societal dependence on natural ecosystems,* Island Press, pp. 93-112.

United Nations Environment Programme 1992. *Convention on biological diversity, June 1992*. Nairobi, United Nations Environment Programme Environmental Law and Institutions Programme Activity Centre.

Whitehead, D. R. (1969). Wind pollination in the angiosperms: evolutionary and environmental considerations, *Evolution,* **23,** 28-35.

Whittaker, R. J. (1995) Disturbed island Ecology. *Trends in Ecology and Evolution* **10,** 421-425.

World Conservation Monitoring Centre (1992) *Global Biodiversity. Status of the earth's living resources,* Chapman & Hall. London.

CHEMODIVERSITY AND THE BIOLOGICAL ROLE OF SECONDARY METABOLITES, SOME THOUGHTS FOR SELECTING PLANT MATERIAL FOR DRUG DEVELOPMENT

R. VERPOORTE
Division of Pharmacognosy, Leiden/Amsterdam Center for Drug Research, Leiden
University, Leiden, The Netherlands
Email: Verpoort@LACDR.LeidenUniv.NL

Abstract

Plants produce various defence compounds; some are constitutively expressed, while others are activated upon cell damage, or are produced after induction through exogenous signalcompounds. From the physiological role of the secondary metabolites in plants one can draw some conclusions about the chance of finding new biologically active compounds. In the search for biologically active compounds, particularly young tissues and seedlings seem to be interesting sources. Treatment of plant material with for example glycosidases may result in the generation of new compounds. Infected plant materials may be an additional source for different types of secondary metabolites, which may also be found in plant cell cultures after treatment with elicitors. Such compounds are particularly of interest for screening for antimicrobial and antitumor activity. Inducible pathways may also be the cause of considerable variation in different collections of plant material.

1. Introduction

Since ancient times people has been exploring nature, and in particular plants, in the search for new drugs. This has resulted in the use of a large number of medicinal plants to treat various diseases. Many of these plants have been shown to be active, and quite a few drugs of western medicine are based on the traditional use of plant drugs. Some are pure compounds from the traditional medicinal plant, others are modifications of these compounds. The traditional uses of medicinal plants remain an important source for drug discovery. However, for finding new leads from plants, at-random screening of plant species is presently widely´used by pharmaceutical industry, as high through-put screening methods allow testing of large number of extracts.

In order to improve the chances to find active compounds various selection criteria can be used to reduce the number of plants to be screened. For example, criteria can be based on chemotaxonomy, traditional use or plant ecological observations. Here I would like to discuss some aspects of the latter approach. Probably it is the oldest, as most likely many of the traditional uses of plants are based upon careful observations made by people, e.g. insecticidal preparations from plants not attacked by insects.

For many years the study of plant secondary metabolism has been neglected as no apparent function was known, secondary metabolites were thought to be waste products of plants. But in the last decade the important role of plant secondary metabolites for among others resistance against pests and diseases has been recognized. How can we use such information to more efficiently find biologically active compounds? Our knowledge in this field is still very limited, and will probably remain limited. With estimations of about 250,000 different plant species, up to 30 million species of insects, 1.5 million species of fungi and similar numbers for algae and prokaryotes (Pimm et al. 1995), the number of possible interactions is almost infinite. During evolution these interactions have resulted in an enormous chemodiversity in nature. The database NAPRALERT contained in 1988 already more than 88,000 secondary metabolites, and every year some 4000 new ones are being reported (N.R. Farnsworth, personal communication), thus there should now be far more than 100,000 secondary metabolites known. This is only from the small percentage of all species studied so far (TABLE 1). Moreover, the species studied were in most cases only studied for a certain type of compounds, e.g. alkaloids, but not further for the presence of others.

TABLE 1. Number of plant species which have been studied with respect to their phytochemistry or for at least one type of biological activity, as on December 1, 1995, in the database NAPRALERT (N.R. Farnsworth, personal communication):

	Biological activity	Phyto-chemically
Monocots	1,283	3,721
Dicots	11,924	31,126
Gymnosperms	239	638
Pteridophytes	349	961
Bryophytes	39	457
Lichens	118	625

In nature chemodiversity is mainly based upon a few building blocks: the phenylpropanoid pathway (C_9-units), the terpenoid or isoprenoid pathway (C_5-units) and the polyketide pathway (C_2-units), and some amino acids. The idea of combinatorial chemistry is in fact as old as evolution. This statement fits with the recent saying that combinatorial chemistry serves as a synthetic rainforest surrogate (Hogan 1997).

Because of our limited knowledge of plant secondary metabolism, it will be difficult to identify a single interaction between a plant and for example an insect for finding a new biologically active compound. However, based on a more general approach one can point out which plant materials are more likely to have high levels of biologically active compounds. In certain stages of plant life or in different plant tissues there will be a higher need for defence compounds than others. There will always be a constitutive level of defence compounds in plants, e.g. antifeedants or any other compound that will help in reducing the level of losses due to pests. Fast growing plants might put more energy in the growth than to spend it into high levels of secondary metabolites. Particularly for young leaves, which have a high nutritional value, the need for defence will be much higher than for old leaves to safeguard the future photosynthetic capacity of the plant. For example Van Dam et al. (Van Dam 1995, Van Dam et al. 1997) showed that in *Cynoglossum officinale* the level of pyrrolizidine alkaloids are highest in young leaves, and that the level declines on aging of the leaves. The maximum level coincides with the maximum of photosynthetic capacity of the leaves.

During the life-cycle of a plant one may thus expect a high level of defence compounds in seeds, germinating seeds and seedlings, as well as in young tissues (e.g. leaves). After attack by insects and/or infection by microorganisms, one may expect that the plants turn on defences against the predator or microorganisms. In this type of defence systems inducible biosynthetic pathways will play a role, which might even be restricted to infected areas only.

The phytoalexin biosynthesis is a good example. In most cases the various defence mechanisms are quite effective, as apparently only a few insect species and microorganisms are able to successfully attack a plant, and thus be a pest or cause a disease. Most pests and pathogenic microorganisms are in fact efficiently controlled by plants.

Below some examples will be given from our own research showing that under certain conditions, there might be better chances of finding interesting secondary metabolites. The examples concern constitutive production of defence compounds, biochemical activation of defence compounds and inducible biosynthesis of defence compounds.

2. Constitutive produced secondary metabolites

Most plants have different types of defences to reduce risks of predation and infection. The defences are in part mechanical, e.g. thorns, hairs, toughness of leaves, and crystals, as well as in part biochemical, e.g. proteases, irritants, antifeedants, and insecticides. The biochemical defences include secondary metabolites already stored in cells, sometimes in specialized cells such as in laticifers, secretory glands and hairs, or stored in specific tissues such as bark and cuticula. Such tissues are thus of great interest for finding biologically active compounds.

One can hypothesize that young tissues and seedlings should be best protected by toxic compounds against predators (see above). This can be illustrated with some results of our studies on the biosynthesis of *Cinchona* alkaloids. Since ancient times, the bark of

Cinchona species was used in South America to treat fever. This bark found its way to Europe, where it became an expensive commodity, eventually resulting in the development of *Cinchona* plantations in Asia. The extracts of the bark were found to be effective in the treatment of malaria. The activity is due to the quinoline alkaloids, which are present in the bark at levels of up to 15% of the dry weight. Quinine, the most active compound, was developed as an antimalaria drug, later its stereoisomer quinidine was developed into a antiarrhythmic drug.

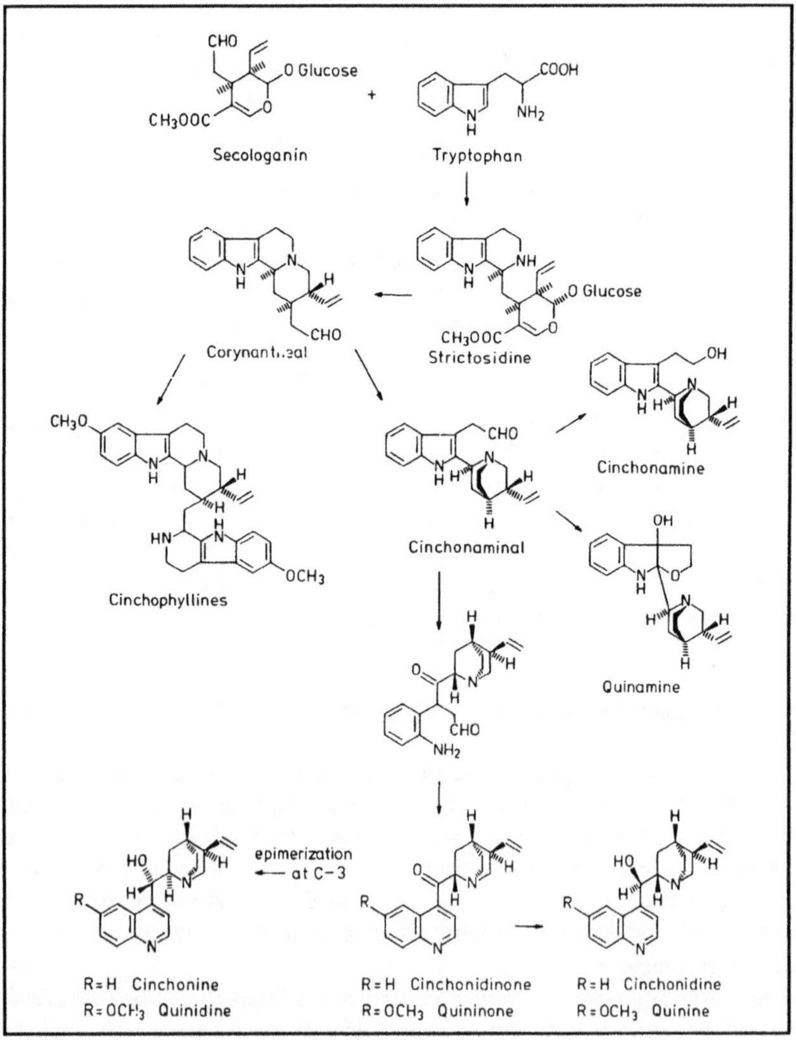

Fig. 1. Biosynthesis of *Cinchona* alkaloids

In our studies of these alkaloids (their biosynthesis is summarized in Fig. 1), we looked for a model system to study the regulation of their biosynthesis. Unfortunately these compounds are only formed in very low levels in *Cinchona* plant cell cultures, which otherwise would have been most suitable for biosynthetic studies. It was then found that during the development from seed to seedling alkaloids were formed. No alkaloids were present in seeds, but after the radicle came through the seedskin (Fig. 2, day 0), the activities of the enzymes of the alkaloid biosynthesis are rapidly induced and quinoline alkaloids are being formed, reaching a maximum level of about 0.4 mM g^{-1} fresh weight of seedling, 4 days later (see Fig. 3) (Aerts et al. 1991a). The increase of the alkaloid level is inversely proportional to the amount of predation by snails (*Deroceras panormytanum*) on the seedlings (Fig. 4). An antifeedant effect of the quinoline alkaloids could be confirmed in experiments with an artificial diet.

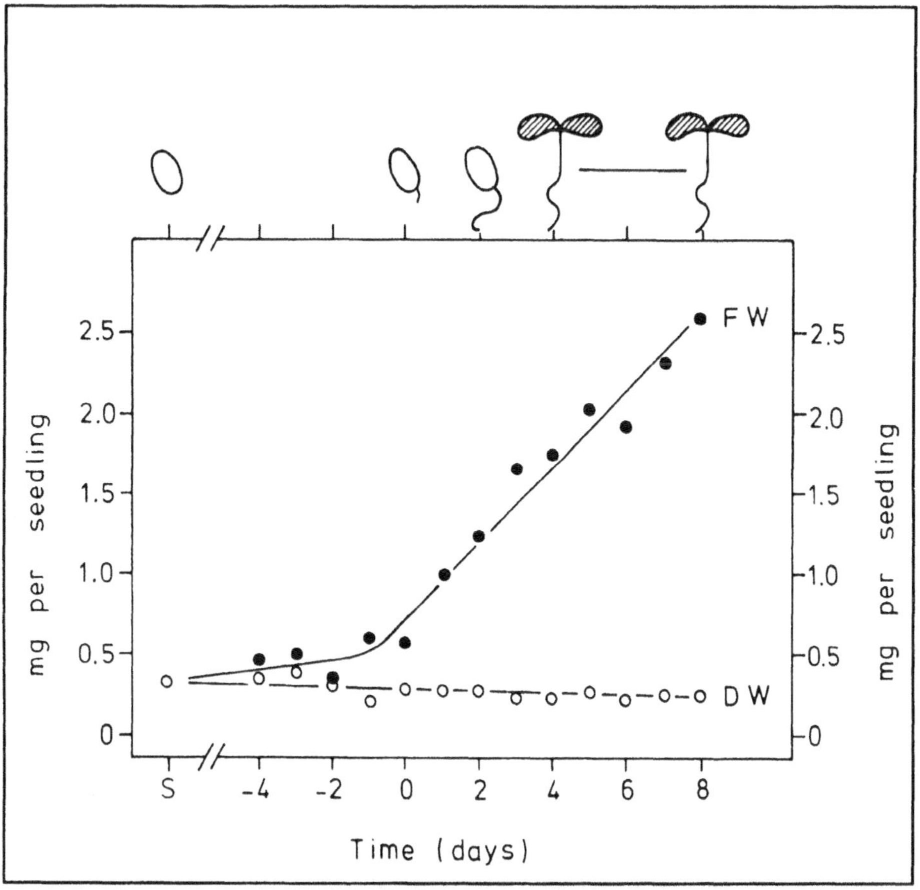

Fig. 2. Development of dry and fresh weight of *Cinchona ledgeriana* seedlings

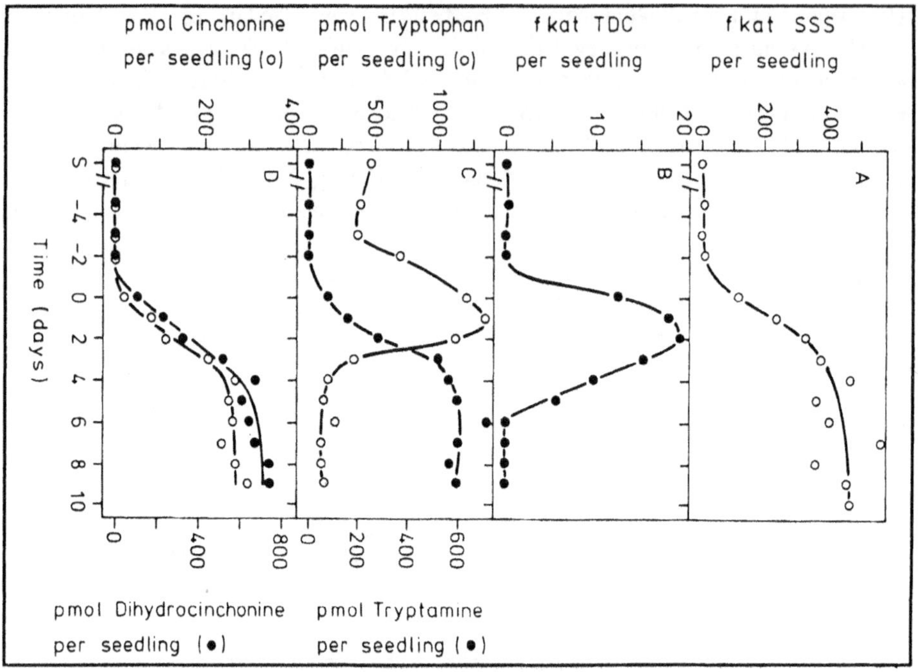

Fig. 3. Activities of some biosynthetic enzymes and alkaloids during germination and growth seedlings *Cinchona ledgeriana*

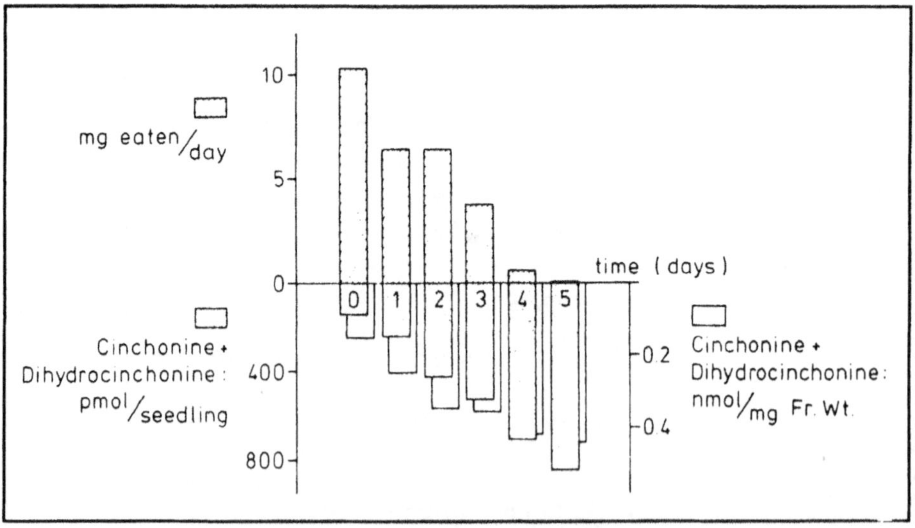

Fig. 4. Amounts eaten by slugs (*Deroceras panormytanum*) at different stages of development of *Cinchona ledgeriana* seedlings

A further illustration of the importance of seedlings as potential source of biologally active compounds is our finding that *Ginkgo biloba* seedlings contain much higher levels of ginkgolides than the mature plant (Carrier et al. 1997). However, one should not conclude that all secondary metabolites from a plant can be found in seedlings. Once more *Cinchona* can serve as example. In further studies of the role of alkaloids in this plant, about 6 month old plants were investigated for the sites of biosynthetic activity and the alkaloids present (Aerts et al. 1991b, 1992). Highest levels of strictosidine synthase were found in young leaves, stem top while somewhat lower levels were found in the lower parts of the stem and the roots (Fig. 5). Alkaloid level is highest in young leaves, however, these alkaloids are of the cinchophylline-type, semidimeric indole alkaloids, also derived form strictosidine, but structurally quite different from the quinolines (Fig. 1). It has been shown that these alkaloids have antimicrobial activity (Caron et al. 1988). We showed that in the plant these compounds serve as antifeedants against larvae, such as *Spodoptora exigua* (Aerts et al. 1991b, 1992). Young leaves contain these alkaloids at levels which give total feeding deterrence, older leaves have much lower levels. It shows thus that plants apparently can make different secondary metabolites from one precursor, serving different purposes in various plant parts.

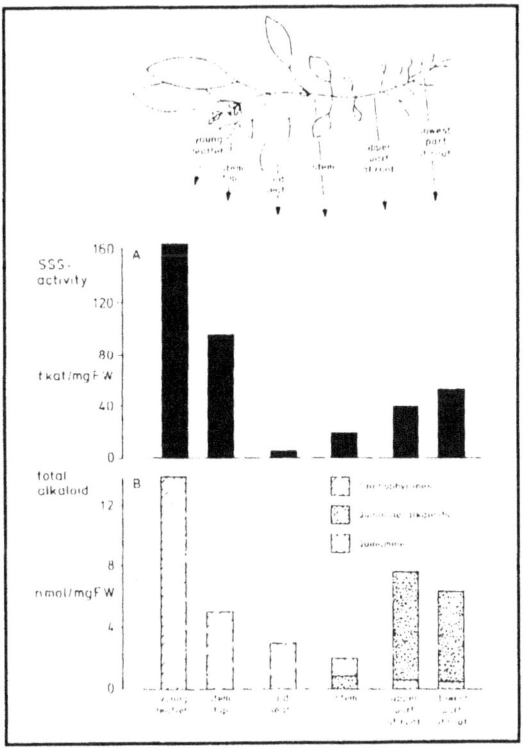

Fig. 5. Levels of strictosidine activity and alkaloids in different parts of 6 month old *Cinchona ledgeriana* plants

Seedlings and young leaves thus seem to be good sources for finding biologically active compounds. However, one should keep in mind that in older leaves other compounds can be found, for example (breakdown) products formed from the major compounds present in the young leaves. In *Catharanthus roseus* young leaves have strictosidine as the major alkaloid, whereas in older leaves this alkaloid is not found anymore. Instead compounds as vindoline, catharanthine, anhydrovinblastine and, as minor compounds, vinblastine and vincristine are present. Thus a further major message is that considerable differences may occur in the spectrum of compounds present, dependant of the age of the harvested plant material.

3. Biochemically activated secondary metabolites

Besides constitutively produced compounds that have a protective effect due to their biological activity, one can also find compounds in plants which in itself have no activity, but which in combination with other factors become active. Probably the best known example are the cyanogenic glycosides, which in combination with a glucosidase yield the highly toxic HCN (Conn 1993). But there are more examples (for a review concerning antifungal compounds see Osbourn 1996). Examples which show that this is a new potential area for leadfinding, at least it shows that the concept of a prodrug could have been learned from nature. Here this will be illustrated with strictosidine, the intermediate in the terpenoid

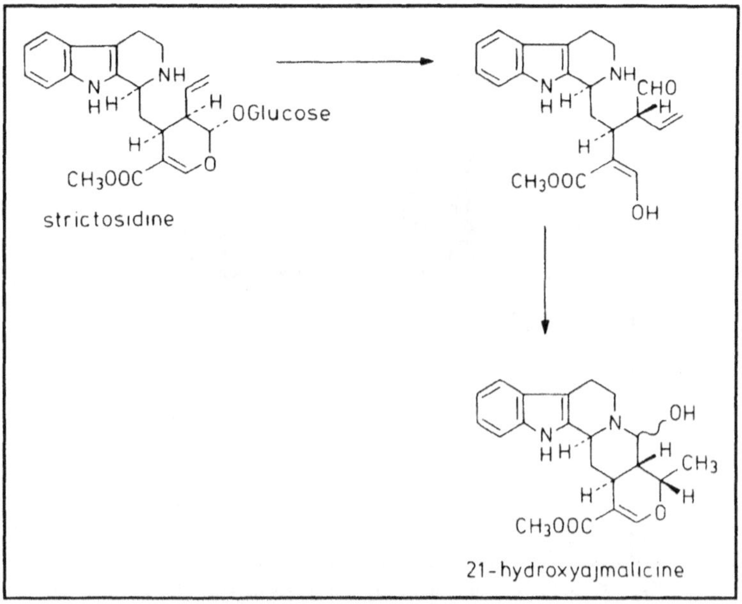

Fig. 6. Glycolysis of strictosidine

indole alkaloid biosynthesis in a number of plant species (Nagakura et al. 1979). Studies on the alkaloids in different parts of *Catharanthus roseus*, showed that young leaves accumulate high levels of strictosidine. From various studies it is known that this gluco-alkaloid is formed and accumulated in the vacuole (Verpoorte et al. 1997). The next step in the biosynthesis of terpenoid indole alkaloids is catalyzed by a specific β-glucosidase. This enzyme, strictosidine glucosidase (SG), splits of the glucosyl moiety. The aglucone has, after ring opening, two reactive aldehyde groups, which can couple with each of the two amine functions in the molecule (Fig. 6). As a result various types of indole alkaloid skeletons can be formed.

SG was recently purified to homogeneity and characterized (Luijendijk 1995, Luijendijk et al. 1996a). This enzyme has a high substrate specifity. Localization studies showed that most of the activity is in the vacuolar fraction of cells from a *C. roseus* cell culture. It is thought that the enzyme is associated with the outside of the tonoplast (Luijendijk 1995, Stevens et al. 1993). For strictosidine itself no antifeedant or antimicrobial activity could be found, but in combination with SG it has strong antimicrobial activity (Luijendijk 1995, Luijendijk et al. 1996b). Such "prodrug-type" phytoanticipines offer another interesting, not yet explored target for screening. Treating plant material with enzymes such as glycosidases before extraction and screening might result in new active compounds not found in non-treated plant material.

4. Biochemically induced secondary metabolism

Upon infection with microorganisms or viruses plants react in a highly complex way. Locally and systemically, different reactions can be observed. Locally by direct induction through signal compounds (elicitors) from the microorganisms, or formed by breakdown of cell walls of the plant. The local induction on the site of infection results in the production of endogenous signal molecules which induce a systemic reaction. Salicylic acid and jasmonic acid are examples of such endogenous signal molecules. The methyl esters of these compounds can even serve as signal compounds between plants (Shulaev et al. 1997). The response to elicitors includes among others the formation of PR-proteins (pathogen-related proteins) (Ryals et al. 1994). These proteins include chitinases and glucanases, which will attack the cell wall of microorganisms. Also secondary metabolism pathways are induced, resulting in the production of phytoalexins. Phytoalexins are defined as antimicrobial active compounds that are synthesized and accumulated by a plant after infection with microorganisms, i.e. in a healthy plant such compounds are not found. Each plant has its own specific set of phytoalexins, which represent many different types of compounds. In several Solanaceae species, such as tobacco, potato and tomato for example sesquiterpenes are formed, whereas in other plants such as peas isoflavones act as phytoalexins (Beerhues 1992, Kuc 1992).

The response of cells to infection can be studied quite well in plant cell cultures. Such cell cultures are excellent model systems to isolate and identify the phytoalexins, and to study their biosynthesis and its regulation. Such elicited cell cultures are an excellent source of antimicrobial active compounds.

The phytoalexin production might be demonstrated by our work on *Cinchona* cell cultures. Cell suspension cultures of different species of this genus have been set up, with the aim to study the production of the pharmaceutically important alkaloids quinine and quinidine in this system. However, these cell cultures produce only trace amounts of these alkaloids. Different media were tested, but none resulted in any significant increase of alkaloid production. But, under certain conditions the cells and the culture medium turned orange-brown. The compounds formed were isolated and identified as anthraquinones (Fig. 7). These compounds had not been reported before from *Cinchona* species, but several other genera in the family of Rubiaceae are wellknown for the production of these coloured compounds, e.g. in the roots of *Rubia tinctorium* and in the flowers, leaves, stem and roots of *Morinda citrifolia* (Wijnsma and Verpoorte 1986c).

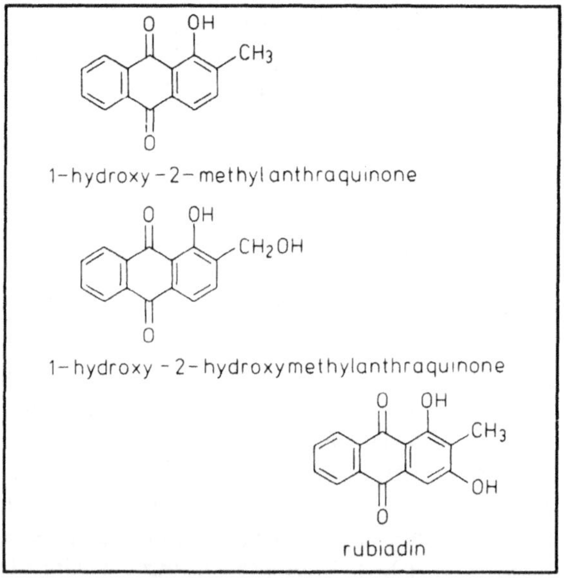

Fig. 7. Some anthraquinones found in *Cinchona* cell cultures

In the *Cinchona* cell cultures the production of the anthraquinones can be induced by various elicitors such as cellulase, pectinase, and cell wall preparations of yeasts and *Phytophthora cinnamomi* (Mulder-Krieger et al. 1982, 1984, Wijnsma et al. 1984, 1985, 1986a). Also jasmonic acid induces the production of these compounds (unpublished results). Further studies showed that these compounds do have antimicrobial activity. To complete the evidence that the anthraquinones are the phytoalexins in *Cinchona* also plants were studied (Wijnsma et al. 1986b). In healthy, 6 months old plants their presence could not be determined. However, after infection with, for example, *Agrobacterium tumefaciens*, the anthraquinones could be detected in the plant. In material collected from trees from plantations in Africa infected with *Phytophthora cinnamomi*, also anthraquinones could be detected. This material had lower alkaloid levels, when compared with healthy parts of the same tree, where no anthraquinones could be detected. This was the final evidence that in *Cinchona* these compounds can be considered as phytoalexins. Interestingly, in the two other Rubiaceae plants mentioned the biosynthesis of the anthraquinones is expressed constitutively, though in different plant parts. The same pathway can thus be regulated quite differently in related plant species. As phytoalexins per definition have antimicrobial activity, they form an interesting target for searching for antibiotics and antitumor compounds. Plant cell cultures induced by jasmonic acid or fungal elicitors should be an excellent source for such a screening. The fact that the phytoalexin biosynthesis is inducible, implies that in collecting plant material one should be aware of the possibility that diseased plants might give different activity profiles than healthy plants. It also means that by just studying healthy plants, one will miss the enormous potential of phytoalexins as a source of biologically active compounds.

5. Conclusion

Here just some examples from our own research were taken, but many others can be found in literature which could also serve to illustrate the points discussed.

From the above examples it is obvious that in young tissues and seedlings interesting biologically active compounds are present at quite high levels. However, there is no indication about the type of activity of these compounds, other than probably an antifeedant or insecticidal effect. Any type of pharmacological activity might be found, like in an at-random screening of plant materials. If we would have screened seedlings of *Ginkgo* and *Cinchona* with the appropriate bioassays (PAF-inhibition, antimalaria, antiarhythmic) we would have found important molecules (ginkgolides, quinine, quinidine). But if we would have screened only young leaves of *C. roseus* we would have missed vinblastine and vincristine. A major advantage of screening seedlings is that it is an easy and reproducible system.

In elicited plant cell cultures, one is likely to find compounds which have antimicrobial activity, compounds which particularly should be of interest to screen for antiviral, antitumor and antimicrobial activity. By screening of the elicited cell cultures of *Cinchona* and *C. roseus* for antimicrobial active compounds we would have found anthraquinones and 2,3-dihydroxybenzoic acid respectively. The anthraquinones might probably also be detected in assays for antitumor or cytotoxic activity.

One interesting point to be noticed is that usually in plants one finds a series of closely related compounds, rather than one single compound. This is similar to the idea of combinatorial chemistry, offering a plant the advantage of having an array of compounds possibly active against an array of organisms. Moreover, development of resistance against a broad range of compounds with similar structures is less likely to occur. Synergism is another phenomenon observed with the mixtures of secondary metabolites (de Jager et al. 1996, Cham and Daunter 1990, Fewell and Roddick 1993, Roddick and Rijnenberg 1986, 1987, Roddick et al. 1988, 1990). This natural concept of mixtures of related compounds could be interesting for drug development, though it would be very much against the present tradition of using single pure chemically entities.

One of the problems in screening plants for biologically active compounds is the variability of the plant material. Besides the well-known variability due to different genotypes, as well as to seasonal and diurnal variation, there will be differences between younger and older parts of plants. Moreover, infected plant might contain completely different compounds, normally not found in the plant.

The final conclusion is that for finding new biologically active compounds, besides screening of plant material as it is widely done nowadays, the screening of seedlings and plant cell cultures is a very interesting option.

6. References

Aerts, R.J., van der Leer, T. van der Heijden, R. and Verpoorte, R. (1990) Developmental regulation of alkaloid production in *Cinchona* seedlings. *J. Plant Physiol.* **136**, 86-91.

Aerts, R.J., Snoeijer, W., Aerts-Teerlink, O., van der Meijden, E., and Verpoorte, R. (1991a) Control and biological implications of alkaloid synthesis in *Cinchona* seedlings. *Phytochemistry* **30**, 3571-3577.

Aerts, R.J., de Waal, A. Pennings, E.J.M., and Verpoorte, R. (1991b) The distribution of strictosidine synthase activity and alkaloids in *Cinchona* plants. *Planta* **183**, 536-541.

Aerts, R.J., Stoker, A., Beishuizen, M., Jaarsma, I., van de Heuvel, M., van der Meijden, E., and Verpoorte, R (1992a) Detrimental effects of *Cinchona* leaf alkaloids on larvae of the polyphagous insect *Spodoptera exigua*. J. *Chem. Ecol.* **18**, 1955-1964.

Beerhues, L. (1992) Pflanzliche Strategien der Pathogenabwehr. *Dtsch. Apoth. Ztg.* **132**, 2485-2491.

Caron, C., Hoizey, M.J., Le Men-Olivier, L., Massiot, G., Zeches, M., Choisy, C., Le Mayre, E., and Verpoorte, R. (1988) Antimicrobial and antifungal activity of quasidimeric and related alkaloids. *Planta Med.* **54**, 409-412.

Carrier, D.J., van Beek, T.A., van der Heijden, R., and Verpoorte, R. (1997) Distribution of ginkgolides and terpenoid biosynthetic activity in *Ginkgo biloba*. Submitted

Cham, B.E. and Daunter, B. (1990) Solasodine glycosides. Selective cytotoxicity for cancer cells and inhibition of cytotoxicity by rhamnose in mice with sarcoma 180. *Cancer Lett.* **55**, 221-225.

Conn, E.E. (1993) β-Glycosidases in plants. In: ACS Symposium Series vol. 533, β-Glycosidases, Biochemistry and Molecular Biology. Esen, E. (Editor). American Chemical Society, Washington DC, pp 15-26.

De Jager, C.M., Butot, R.P.T., van der Meijden, E., and Verpoorte, R. (1996) The role of primary- and secondary metabolites in chrysanthemum resistance to *Frankliniella occidentalis*. *J. Chem. Ecol.* **22**, 1987-1999

Fewell, A.M., and Roddick, J.G. (1993) Interactive antifungal activity of the glycoalkaloids α-solanine and chaconine, *Phytochemistry* **33**, 323-328.

Hogan, J.C. (1997) Combinatorial chemistry in drug discovery. *Nature Biotechnol.* 15, 328-330.

Kuc, J. (1992) Antifungal compounds from plants. In: Phytochemical resources for medicine and agriculture. (Nigg, H.N. and Seigler, D., Eds). Plenum Press New York 1992, pp 159-184.

Luijendijk, T.J.C. (1995) Strictosidine glucosidase in indole alkaloid biosynthesis. PhD-Thesis, University of Leiden.

Luijendijk, T.J.C., Nowak, A., and Verpoorte, R. (1996a) Strictosidine glucosidase from suspension cultured cells of *Tabernaemontana divaricata*. *Phytochemistry* **41**, 1451-1456.

Luijendijk, T.J.C., van der Meijden, E., and Verpoorte, R. (1996b) Involvement of strictosidine as a defensive chemical in *Catharanthus roseus*. *J. Chem. Ecol.* **22**, 1355-1366.

Mulder-Krieger, Th., Verpoorte, R., de Water, A., van Gessel, M., van Oeveren, B.C.J.A., and Baerheim Svendsen, A. (1982) Identification of the alkaloids and anthraquinones in *Cinchona ledgeriana* Moens unorganized callus cultures. *Planta Med.*46, 19-24

Mulder-Krieger, Th., Verpoorte, R., van der Kreek, M., and Baerheim Svendsen A. (1984) Identification of alkaloids and anthraquinones in *Cinchona pubescens* callus cultures; the effect of plant growth regulators and light on the alkaloid content. *Planta Med.* **50**, 17-20.

Nagakura, N., Rueffer, M., and Zenk, M.H. (1979) The biosynthesis of monoterpenoid indole alkaloids fromstrictosidine. *J. Chem. Soc (Perkin I)* 2308-2312.

Osbourn, A.E. (1996) Preformed antimicrobial compounds and plant defense against fungal attack. *Plant Cell* **8**, 1821-1831.

Ryals, J., Uknes, S., and Ward, E. (1994) Systemic acquired resistance.*Plant Physiol.* **104**, 1109-1112.

Pimm, S.L., Russell, G.J., Gittleman J.L. and Brooks, T.M. (1995) The future of biodiversity. *Science* **269**, 347-350.

Roddick, J.G., and Rijnenberg, A.L. (1986) Effect of steroidal glycoalkaloids of the potato on the permeability of liposome membranes, *Physiol. Plant* **68**, 436-440.

Roddick, J.G. (1987) Antifungal activity of plant steroids. In: Lipid interactions among plants and their pests and pathogens, Fuller, G., and Nes, W.D. (Editors), ACS Symposium Series 325, American Chemical Society, Washington, pp 201-222.

Roddick, J.G., and Rijnenberg, A.L. (1987) Synergistic interaction between the potato glycoalkaloids α-solanine and α-chaconine in relation to lysis of phospholipid/sterol liposomes, *Phytochemistry* **26**, 1325-1328.

Roddick, J.G., Rijnenberg, A.L., and Osman, S.F. (1988) Synergistic interaction between potato glycoalkaloids α-solanine and α-chaconine in relation to destabilization of cell membranes: ecological implications, *J. Chem. Ecol* **14**, 889-902.

Roddick, J.G., Rijnenberg, A.L., and Weissenberg (1990) Membrane-disrupting properties of the steroidal glycoalkaloids solasonine and solamargine, *Phytochemistry* **29**, 1513-1518.

Shulaev, V., Silverman, P., and Raskin, I. (1997) Airborne signalling by methyl salicylate in plant pathogen resistance *Nature* **385**, 718-721.

Stevens, L.H., Blom, Th.J., and Verpoorte, R. (1993) Subcellular localization of tryptophan decarboxylase, strictosidine synthase and strictosidine glucosidase in cell suspension cultures of *Catharanthus roseus* and *Tabernaemontana divaricata*. *Plant Cell Rep.* **12**, 573-577.

van Dam, N.M. (1995) Production, distribution and function of secondary metabolites. A case study on pyrrolizidine alkaloids in *Cynoglossum officinale*. PhD Thesis, Leiden University.

Van Dam, N.M., de Jong, T.J., Iwasa, Y., and Kubo, T. (1997) Optimum defence distribution within rosette plants of *Cynoglossum officinale* L.: are plants smart investors? *Functional Ecol.* in press.

Verpoorte, R. van der Heijden, R., and Moreno, P.R.H. (1997) Biosynthesis of terpenoid indole alkaloids in *Catharanthus roseus*. The Alkaloids. G.A. Cordell (Editor) Academic Press, San Diego. In press.

Wijnsma, R., Verpoorte, R., Mulder-Krieger, Th., and Baerheim Svendsen, A. (1984) Anthraquinones from callus cultures of *Cinchona ledgeriana*. *Phytochemistry* **23**, 2307-2311

Wijnsma, R., Go, J.T.K.A., van Weerden, I.N., Harkes, P.A.A., Verpoorte, R., and Baerheim Svendsen, A. (1985) Anthraquinones as phytoalexines in cell and tissue cultures of *Cinchona* species. *Plant Cell Rep.* **4**, 241-244.

Wijnsma, R., Go, J.T.K.A., Harkes, P.A.A., Verporte, R., and A. Baerheim Svendsen, A. (1986a) Anthraquinones in callus cultures of *Cinchona pubescens*. *Phytochemistry* **25**, 1123-1126.

Wijnsma, R., van Weerden, I.N., Verpoorte, R., Harkes, P.A.A., Lugt, Ch.B., Scheffer, J.J.C., and Baerheim Svendsen, A. (1986b) Anthraquinones in *Cinchona ledgeriana* bark infected with *Phytophthora cinnamomi*. Planta Med. **52**, 211-212.

Wijnsma, R., and Verpoorte, R. (1986c) Anthraquinones in Rubiaceae. *Fortschr. Chem. Organ. Naturstoffe* **49**, 79-149.

EVOLVING STRATEGIES FOR THE SELECTION, DEREPLICATION AND PRIORITIZATION OF ANTITUMOR AND HIV-INHIBITORY NATURAL PRODUCTS EXTRACTS

JOHN H. CARDELLINA II, RICHARD W. FULLER,
WILLIAM R. GAMBLE, CHANDRA WESTERGAARD,*
JAMIE BOSWELL,* MURRAY H. G. MUNRO,**
MICHAEL CURRENS AND MICHAEL R. BOYD
Laboratory of Drug Discovery Research and Development, Developmental Therapeutics Program,
Division of Cancer Treatment and Diagnosis,
National Cancer Institute,
Frederick Cancer Research and Development Center
P.O. Box B, Bldg. 1052, Room 121,
Frederick, MD 21702-1201

1. Introduction

There is ever increasing pressure in all pharmaceutical lead discovery laboratories and programs to identify novel bioactive chemical entities more quickly and efficiently. In order to accomplish this, the process of selecting which screening leads to pursue or advance has taken center stage as the most critical task in the discovery process. In programs emphasizing natural products as a source of such lead compounds, this task is complicated by the need to differentiate activity due to: a) commonly recurring classes of "nuisance compounds", so-called because they provide a non-specific response in a large number of bioassays; b) known secondary metabolites which may or may not occur widely, but which provide a specific response in the assay in question; and c) novel chemical entities which are specifically active in the target assay.

Dereplication is a term used to describe efforts to identify or characterize, with speed and efficiency, leads of types "a" or "b" above. Examples of nuisance natural products include polyphenolics (tannins in terrestrial plants and polyphloroglucinols in brown algae) [1] and alkylated pyridinium polymers from marine sponges. Such compounds typically give a positive response in a broad range of *in vitro* bioassays, resulting in abnormally

* Werner H. Kirsten High School Interns, 1995-6 (CW) and 1996-97 (JB)
** On leave from University of Canterbury, Christchurch, New Zealand

high hit rates and complicating the process of identifying or selecting viable lead extracts for bioassay-guided fractionation. Our decade of experience in searching for HIV-inhibitory natural products has led us to recognize that other recurring compounds, some of more limited distribution, may interfere in specific assays. Examples include anionic polysaccharides [2] (broad distribution), sulfated sterols (sponges, echinoderms, hydroids) [3] and sulfolipids (photosynthetic organisms, particularly photosynthetic microorganisms) [4]. We have adapted protocols for the routine dereplication of anionic polysaccharides [2] and polyphenolics [1] from screening leads prior to selection for project status, but frequently find ourselves reisolating known compounds with known bioactivity from sources similar to [5,6] or different from [7,8] those previously reported. The isolation of a previously known compound of limited distribution from a new source is a particularly vexing problem in natural products; one simply can not anticipate the dereplication challenge. This problem has had an increasing impact in many laboratories, including ours. For example, we have recently expended considerable effort isolating, purifying and identifying two marine cytotoxins from new genus sources: jaspamide [9] (**1**, also known as jasplakinolide [10]), from the sponge *Dorypleres splendens* (Fuller et al, LDDRD, unpublished) and hemiasterlin [11,12] (**2**), from the sponge *Siphonochalina* sp. (Gamble et al, LDDRD, manuscript in preparation). Such experiences have prompted us to examine additional strategies to improve the process of selection, dereplication and prioritization.

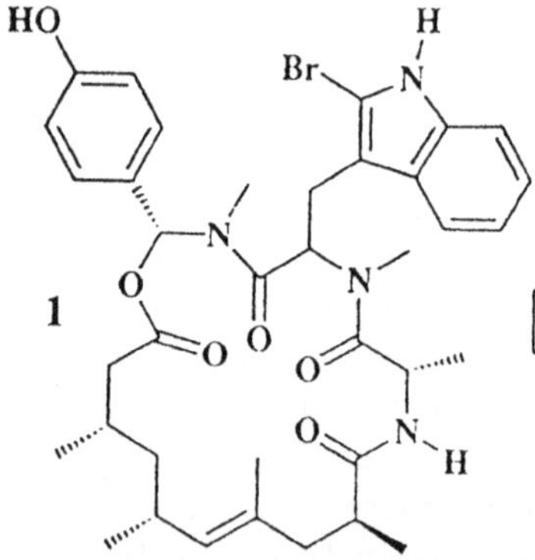

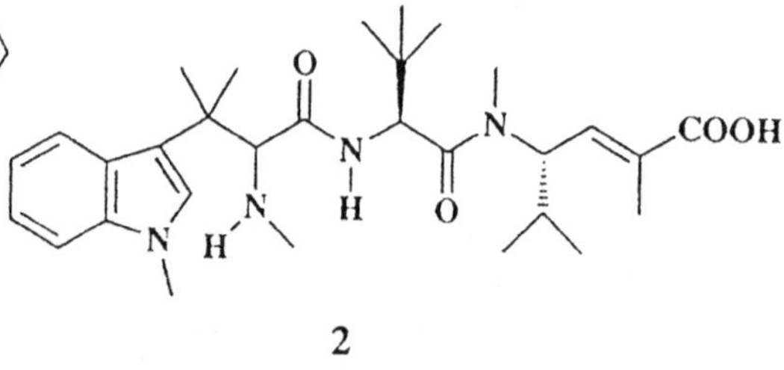

2. Chemotaxonomy in Dereplication

Taxonomy can be a powerful tool in dereplication and prioritization, especially in cases where secondary metabolites have a rather limited distribution, such as restriction to one or a few families. The large plant family Guttiferae (Clusiaceae) has previously yielded two very distinct anti-HIV compound classes, the calanolides [13-16] and the guttiferones [17]. The NCI Repository contains a considerable number of collections from this family, and extracts from this group have shown somewhat higher than average positive responses in the anti-HIV primary screen. As a consequence, we felt it necessary to incorporate dereplication steps for these two compound classes in our preliminary evaluation and prioritization of screening leads from the Guttiferae.

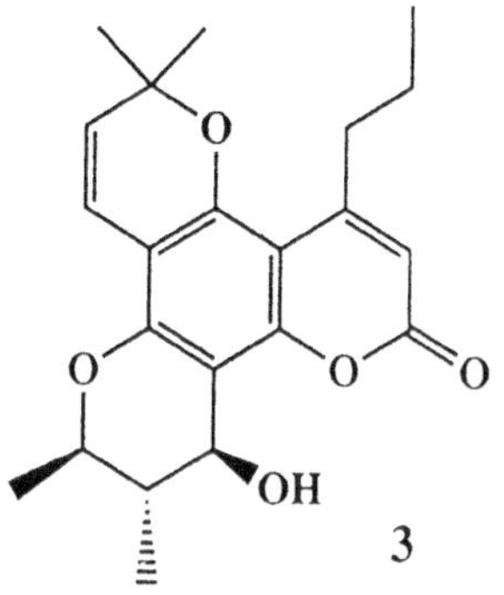
3

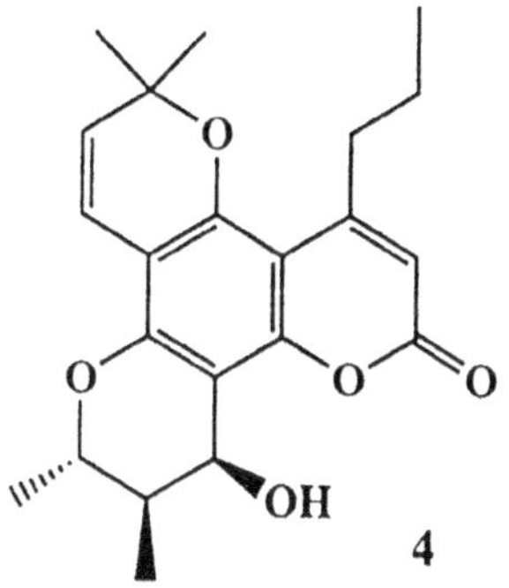
4

The calanolides are members of the class of prenylated coumarins, known for over four decades to occur in the genus *Calophyllum* [18]. (+)-Calanolide A (**3**) and (-)-calanolide B (**4**) were isolated in rather low yields from the leaves and twigs of *C. lanigerum* var. *austrocoriaceum* following anti-HIV bioassay-guided fractionation. They are potent blockers of the cytopathic effects of HIV-1 *in vitro* [14] and specifically inhibit HIV-1 reverse transcriptase [14-16]. The calanolide class interacts with HIV-1 RT in a manner different from virtually all other known strains of HIV-1. A chemotaxonomic survey of the NCI Repository collections of *Calophyllum* has been undertaken to search for a more abundant source of **3** and/or **4** and to identify additional, hopefully antiviral, members of this compound class. Only 9 of 31 species of *Calophyllum* examined contained prenylated coumarins related to the calanolides. All other antiviral activity was tracked to polyphenolic (tannin) fractions (McKee et al, LDDRD, manuscript in preparation).

Several years ago, LDDRD scientists also isolated a series of denselyprenylated benzophenones as HIV-inhibitory constituents of extracts from the genera *Garcinia*, *Clusia* and *Symphonia* [14]. Those compounds, the guttiferones (see, for example, guttiferone E, **5**), were new members of the garcinol family. Unfortunately, the available guttiferones lacked sufficient potency or *in vitro* therapeutic indices for consideration as potential drug development candidates. As a consequence, they became an object of dereplication efforts by our laboratory for extracts from this family.

A selection of 62 HIV-inhibitory organic extracts, representing 11 genera from the family Guttiferae, was assembled for this study. These extracts were examined by thin-layer chromatography, using guttiferone A and calanolide A as reference standards. None of the extracts (other than certain *Calophyllum* spp.) produced spots with the characteristic R_f values or deep blue staining of the calanolides; on the other hand, several extracts gave

TLC evidence suggestive of, but not conclusive for, the presence of guttiferones (Table 1). Extracts from the genera *Garcinia, Clusia* and *Symphonia* which gave indication of the presence of guttiferones were excluded from further investigation. Coupled with dereplication for tannins, this approach eliminated a high percentage of these leads and focused our attention on a selected few.

TABLE 1. Chemotaxonomic survey of Guttiferae/Clusiaceae for Guttiferone/Garcinol Prenylated Benzophenones[a]

Genus	# species examined	# species positive[b]
Allanblackia	1	1
Calophyllum	26	0
Clusia	9	5
Eliaea	1	1
Garcinia	10	8
Mammea	3	0
Marila	1	1
Psorospermum	1	0
Symphonia	3	2
Tovomita	4	0
Vismia	3	1

[a] TLC analysis, with guttiferone A [14] as standard

[b] R_f values, UV and staining characteristics (vanillin-H_2SO_4) similar to guttiferone A.

The antiviral activity in one of those leads, an extract of roots of *Allanblackia stuhlmannii*, was rapidly traced to a single constituent, a new member of the guttiferone class, **6**. Like the previously identified members of this class, guttiferone F (**6**) exhibited anti-HIV activity, but was also quite toxic and did not fully protect against the cytopathic effects of the virus. The genus *Allanblackia* was not previously known as a source of the garcinol/ guttiferone class.

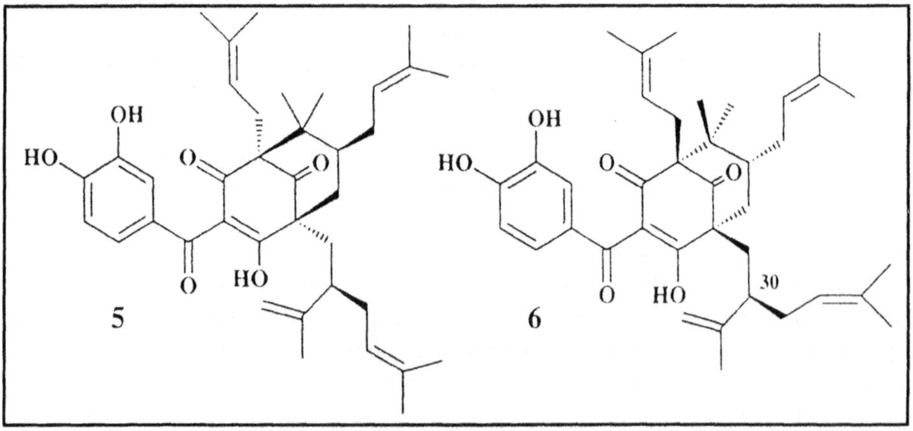

5

6

An extract from leaves of *Vismia cayennensis* also gave a TLC response suggestive of the guttiferones. Bioassay-guided fractionation of the organic extract by solvent partitioning, gel permeation chromatography (Sephadex LH-20), and HPLC led to four benzophenones, **7** - **10**, related to the vismiaphenones [19, 20]. Of these, only **7** exhibited any anti-HIV activity, but with incomplete cytoprotection and only a narrow *in vitro* therapeutic index (EC_{50} ~ 11μg/mL; IC_{50} ~ 30μg/mL), much like the guttiferones. Previously isolated vismiaphenones [19,20] lacked the *p*-hydroxy substituent on the non-prenylated aromatic ring; this might account for the lack of antiviral activity in other examined *Vismia* collections, but has not been verified. The vismiaphenones could be construed as biosynthetic relatives of the guttiferones, lacking the same oxidation pattern on the non-prenylated ring and the more extensive prenylation and carbocylization observed in the guttiferones.

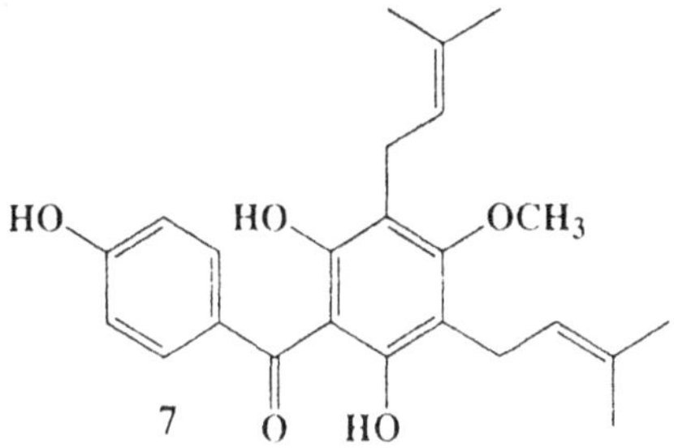

7

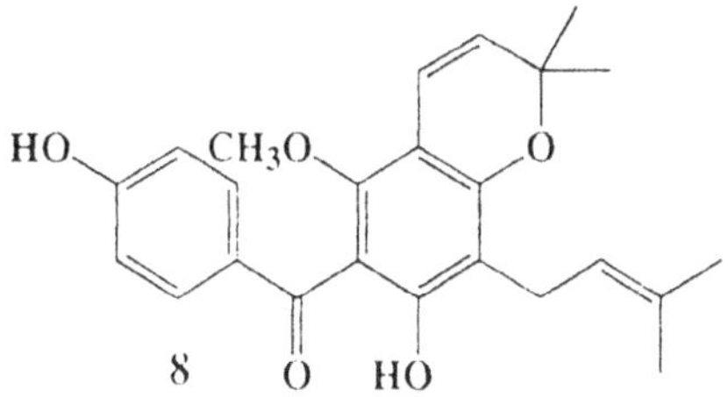

8

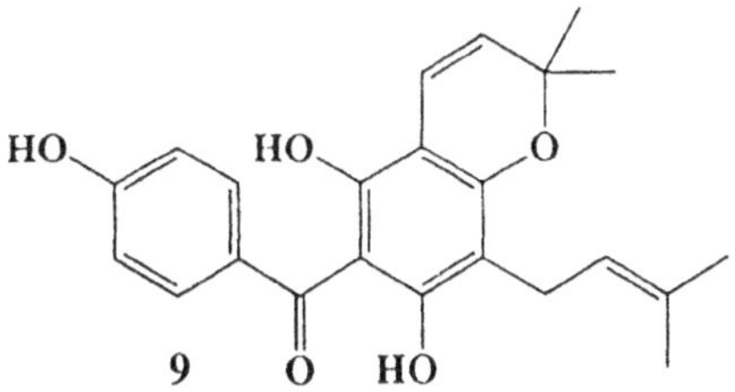

9

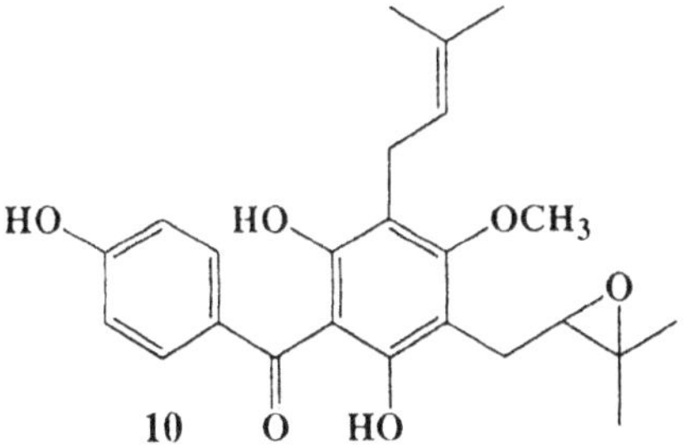

10

These rather extensive follow-up studies of representatives of the Guttiferae have thus far failed to identify any additional new anti-HIV chemotypes from this family. Instead, the results pointed only to polyphenolics, prenylated benzophenones and the calanolide-type coumarins as HIV-inhibitory constituents in this family. Distribution of the calanolides would appear to be limited to relatively few species of the genus *Calophyllum*, while the prenylated benzophenones are more widely distributed in the family, and certainly prevalent in *Garcinia*, *Symphonia* and *Clusia*. It will be interesting to determine whether any unique variants are to be found in the genera *Eliaea* and *Marila* (see Table 1).

3. HPLC, Coupled to Multiple Analytical Methods

Labor intensive protocols, like the TLC study described above, are incompatible with increasing pressures on the time and resources available for the dereplication and prioritization of leads. As a consequence, we have been exploring the concept of a single analytical procedure (HPLC) that would yield multiple data sets (retention times, UV spectra, mass spectra and bioassays). Originally introduced to our laboratory as a more sophisticated version of our chemical screening protocol [1] (by co-author MHGM), the power of the method is amplified by inclusion of a bioassay step.

Briefly, a solvent-solvent partition or solid-phase extraction fraction from an antitumor lead extract is subjected to analytical HPLC (C_{18} column, gradient elution with CH_3CN-H_2O containing TFA). The eluant passes through a diode array (UV-visible) detector and is collected by time increments in a 96-well microtitre plate. Daughter plates can be prepared from this collection plate by transferring aliquots directly or by serial dilution, as warranted. One daughter plate is assayed for cytotoxicity to identify fractions with potential antitumor activity. The corresponding wells in the mother plate plate are then subjected to electrospray mass spectrometry to determine the molecular weight of the active constituent(s). Then, the molecular weight and UV data, and any available taxonomic information, can be used to search natural product databases for potential matches. Two recentexamples from our laboratory illustrate the utility of this approach. In one case, taxonomy and the potency of the extracts clearly directed our attention to a known compound, while the second case illustrates a truly blind dereplication.

Collections of the zoanthid genus *Palythoa* have been made from various geographic locations over the past decade by NCI contractors. The aqueous extracts of a significant number of *Palythoa* collections have exhibited markedly potent cytotoxicity in the NCI 60-cell *in vitro* screen. Previous efforts by our group to determine whether palytoxin [21,22] was responsible for the observed cytotoxicity were unsuccessful, primarily because of the miniscule quantities of the toxin typically present in the zoanthid. Therefore, we applied our evolving multifaceted analytical approach to this problem.

First, a commercial sample of palytoxin was used as a reference standard, in order to establish retention time and detection limits (UV, bioassay and MS). Remarkably, a 5μg sample of palytoxin provided adequate UV and MS signals, but required a 1000-fold dilution to avoid saturating the cytotoxicity assay. Then, crude aqueous extracts of several *Palythoa* collections were subjected to EtOH precipitation to remove polysaccharides and some large proteins. The supernatants were then extracted with *n*BuOH, followed by solid-phase extraction on C_4 media with an H_2O-MeOH step gradient. The 2:1 MeOH-H_2O fractions were subjected to our analytical protocol. In all cases, bioactivity was associated with fractions of similar retention time and identical UV spectra, relative to palytoxin (Figure 1). Mass spectral analyses indicated a series of compounds of similar mass (representing more or less oxygenation, or one to a few additional or fewer carbon atoms), suggestive of the presence of the ostreocins [23] or related compounds; there was also indication of a series of possibly truncated palytoxins (100-150 daltons smaller than palytoxin), which could be of considerable scientific interest.

In the second case, an organic extract of the sponge *Axinella aplysinoides* cf. exhibited potent and differential cytotoxicity in the NCI primary antitumor screen. Solvent-solvent partitioning concentrated the activity in the methyl *t*-butyl ether (MeOtBu) fraction; NMR analysis indicated a rather uncomplicated, nearly pure sample. Analysis of this fraction by HPLC gave a relatively simple chromatogram (see Figure 2) dominated by two peaks with identical UV spectra.

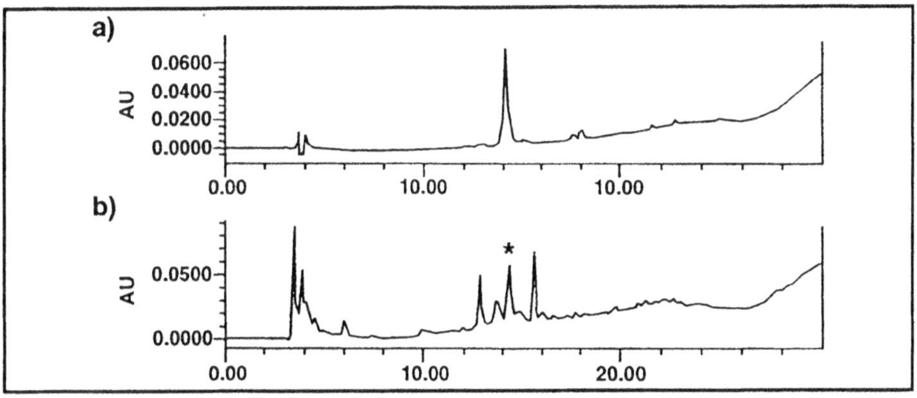

Figure 1. HPLC Chromatograms (233 nm) of : a) palytoxin standard and b) aqueous extract of *Palythoa* sp. (C14974). In (b), peak marked with * is primary cytotoxin.

Bioactivity was concentrated in fractions representing those peaks. FAB mass spectral analyses provided strong pseudomolecular ions (Figure 3). A search of the MarinLit database with those molecular weights yielded only one match for each. Subsequent comparison of the UV and ^{1}H NMR spectra confirmed the identity of the two principal cytotoxins as kabiramides B and C (**11**, **12**) [24,25].

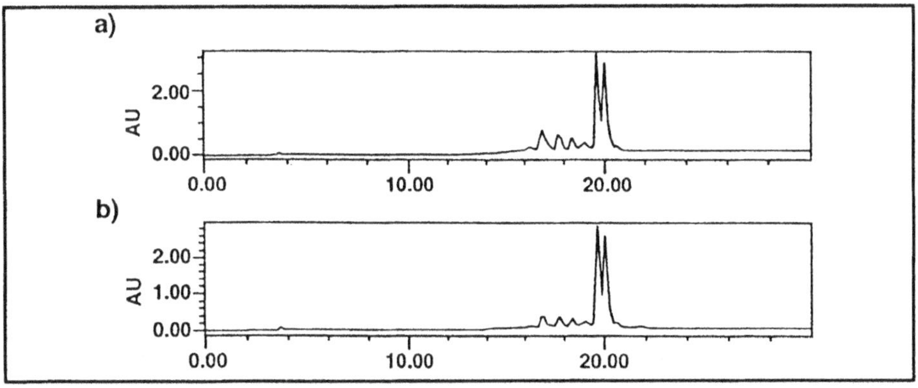

Figure 2. HPLC Chromatograms of MTBE Solubles from Organic Extract of *Axinella aplysinoides* cf.: a) 220 nm; b) 280 nm. The two major peaks (t_R ~ 19.75 and 20 min) are the major cytotoxins.

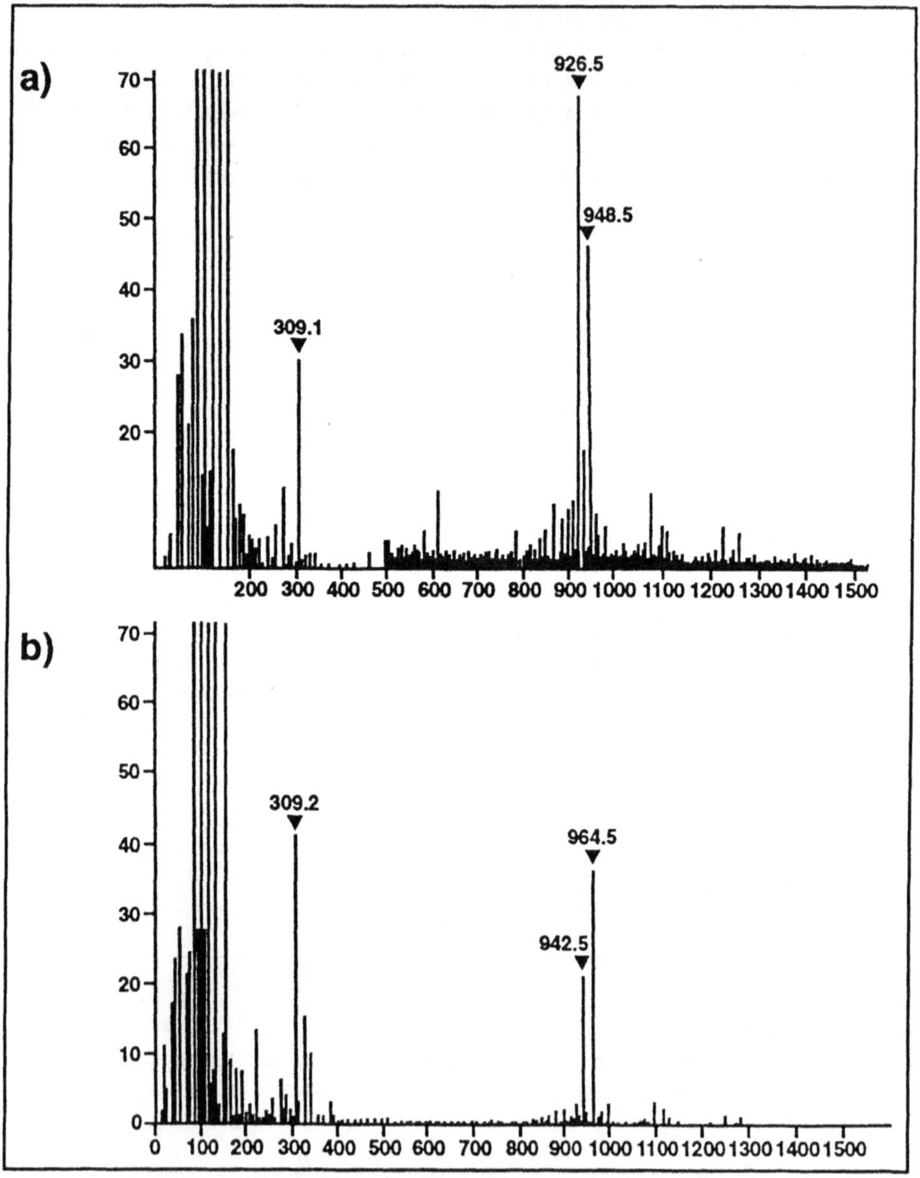

Figure 3. FAB Mass Spectra of Cytotoxic Fractions from *Axinella aplysinoides* cf. (see Figure 2): a) t_R ~ 19.75 min.; b) t_R ~ 20 min.

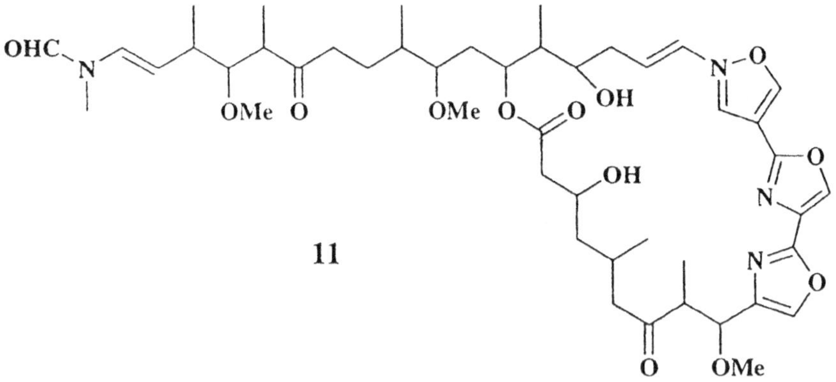

11

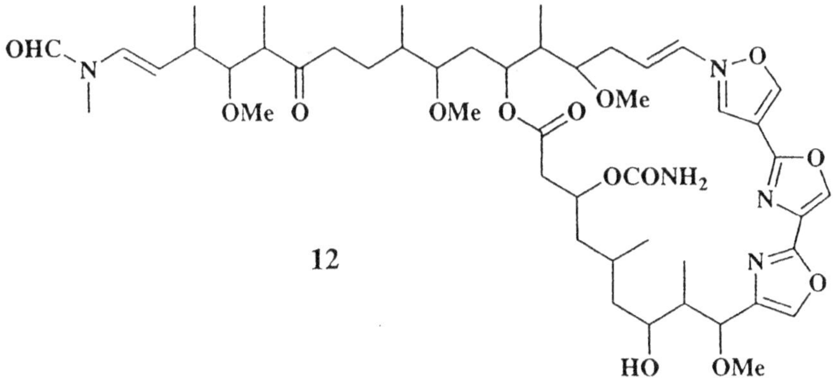

12

4. Discussion

We have concluded that there is no single or universal scheme for dereplication. Some decisions or choices would appear to be obvious. For example, if one were to screen plant extracts for virtually any bioactivity *in vitro*, a dereplication for tannins should be presumed necessary. Likewise, a study of aqueous extracts should probably include dereplication of anionic polysaccharides. It is also apparent that dereplication strategies must be dynamic and flexible, constantly modified as experience is gained with a particular screen or bioassay and additional recurring classes of bioactive compounds are identified.

Our evolving multi-hyphenated HPLC analysis (currently HPLC-UV-MS-bioassay) potentially offers considerable savings in the time and resources allocated to lead identification. The bioassay data limits the number of fractions (or microtitre plate wells) to be analyzed, and the combination of UV and mass spectral data make firm identification of constituents more feasible. If a reference standard of a suspected known compound were available, a third point of comparison (retentiontime) would offer final proof or confirmation of identity.

Increased attention to developing streamlined, efficient, reliable methods of dereplication will permit natural products chemists to focus more attention and effort to isolating and identifying truly novel bioactive lead compounds, the ultimate goal of such a program.

5. References

1. Cardellina, J.H., II, Munro, M.H.G., Fuller, R.W., Manfredi, K.P., McKee, T.C., Tischler, M., Bokesch, H.R., Gustafson K.R., Beutler, J.A. and Boyd, M.R.: A chemical screening strategy for the dereplication and prioritization of HIV-inhibitory aqueous natural products extracts. *J. Nat. Prod.* 56: 1123-1129, 1993.
2. Beutler, J.A., McKee, T.C., Fuller, R.W., Tischler, M., Cardellina, J.H., II, McCloud, T.G., Snader, K.M. and Boyd, M.R.: Frequent occurrence of HIV-inhibitory sulfated polysaccharides from marine invertebrates. *Antiviral Chem. Chemother.* 3: 167-172, 1993.
3. McKee, T.C., Cardellina, J.H., II, Riccio, R., D'Auria, M.V., Iorizzi, M., Minale, L., Moran, R.A., Gulakowski, R.J., McMahon, J.B., Buckheit, R.W., Jr., Snader, K.M. and Boyd, M.R.: HIV inhibitory natural products. 11. Comparative studies of sulfated sterols from marine invertebrates. *J. Med. Chem.* 37: 793-797, 1994.
4. Gustafson, K.R., Cardellina, J.H., II, Fuller, R.W., Weislow, O.S., Kiser, R.F., Snader, K.M., Patterson, G.M.L. and Boyd, M.R.: AIDS-antiviral sulfolipids from cyanobacteria (blue-green algae). *J. Natl. Cancer Inst.* 16: 1254-1258, 1989.
5. Tischler, M., Cardellina, J.H., II, Cragg, G.M. and Boyd, M.R.: Cytotoxic quassinoids from *Cedronia granatensis* Cautrec. *J. Nat. Prod.* 55: 667-671, 1992.
6. Decosterd, L., Gustafson, K.R., Cardellina, J.H., II, Cragg, G.M. and Boyd, M.R.: The differential cytotoxicity of cardenolides from *Thevetia ahouia*. *Phytother. Res.* 8: 74-77, 1994.
7. Fuller, R.W., Cardellina, J.H., II, Cragg, G.M. and Boyd, M.R.: Cucurbitacins: differential cytotoxicity, dereplication and first isolation from *Gonystylus keithii*. *J. Nat. Prod.* 57: 1442-1445, 1994.
8. Rashid, M.R., Gustafson, K.R., Cardellina, J.H., II and Boyd, M.R.: Mycalolides D and E, new cytotoxic macrolides from a collection of the stony coral *Tubastrea* sp. *J. Nat. Prod.* 58: 1120-1125, 1995.
9. Zabriskie, T.M., Klocke, J.A., Ireland, C.M., Marcus, A.H., Molinski, T.F., Faulkner, D.J.,Xu, C. and Clardy, J.C.: Jaspamide, a modified peptide from a *Jaspis* sponge, with insecticidal and antifungal activity. *J. Am. Chem. Soc.* 108: 3123-3124, 1986.
10. Crews, P., Manes, L.V. and Boehler, M.: Jasplakinolide, a cyclodepsipeptide from the marine sponge *Jaspis* sp. *Tetrahedron Lett.* 27: 2797-2800, 1986.
11. Talpir, R., Benayahu, Y., Kashman, Y., Pannell, L. and Schleyer, M.: Hemiasterlin and geodiamolide TA; Two new cytotoxic peptides from the marine sponge *Hemiasterella minor* (Kirkpatrick). *Tetrahedron Lett.* 35: 4453-4453, 1994.
12. Coleman, J.E., deSilva, E.D., Kong, F. and Andersen, R.J.: Cytotoxic peptides from the marine sponge *Cymbastela* sp. *Tetrahedron* 51: 10653-10662, 1995.
13. Kashman, Y., Gustafson, K.R., Fuller, R.W., Cardellina, J.H., II, McMahon, J.B., Currens, M.J., Buckheit, R.W., Jr., Hughes, S.H., Cragg, G.M. and Boyd, M.R.: The calanolides, novel HIV-inhibitory coumarins from the tropical rainforest tree *Calophyllum lanigerum*. *J. Med. Chem.* 35: 2735-2742, 1992.
14. Currens, M.J., Gulakowski, R.J., Mariner, J.M., Moran, R.A., Buckheit, R.W., Gustafson, K.R., McMahon, J.B. and Boyd, M.R.: Antiviral activity and mechanism of action of calanolide a against the Human

Immunodeficiency Virus Type-1. *J. Pharmacol. Exp. Ther.* 279: 645-651, 1996.
15. Hizi, A., Tal, R., Shaharabany, M., Currens, M.J., Boyd, M.R., Hughes, S.B. and McMahon, J.B.: Specific inhibition of the reverse transcriptase of Human Immunodeficiency Virus Type 1 and the chimeric enzymes of Human Immunodeficiency Virus Type 1 and Type 2 by nonnucleoside inhibitors. *Antimicrob. Agents Chemother.* 37: 1037-1042, 1993.
16. Currens, M.J., Mariner, J.M., McMahon, J.B. and Boyd, M.R.: Kinetic analysis of inhibition of Human Immunodeficiency Virus Type-1 reverse transcriptase by calanolide A. *J. Pharmacol. Exp. Ther.* 279: 652-661, 1996.
17. Gustafson, K.R., Blunt, J.W., Munro, M.H.G., Fuller, R.W., McKee, T.C., Cardellina, J.H., II, McMahon, J.B., Cragg, G.M. and Boyd, M.R.: The guttiferones, HIV-inhibitory benzophenones from *Symphonia globulifera, Garcinia livingstonei, Garcinia ovalifolia* and *Clusia rosea*. *Tetrahedron* 48: 10093-10102, 1992.
18. Ormancey-Poitier, A., Buzas, A. and Lederer, E.: Calophyllolide and calophylic acid from the seeds of *Calophyllum inophyllum*. *Bull. Soc. Chem.* Fr. 577-580, 1951.
19. Monache, G.D., Gonzalez, J.G., Monache, F.D. and Bettolo, G.B.M.: Prenylated benzophenones from *Vismia decipiens*. *Phytochemistry* 19: 2025-2028, 1980.
20. Pathak, V.P., Khanna, R.N.: Synthesis of naturally occurring prenylated benzophenones. Vismiaphenone A, vismiaphenone B and isovismiaphenone B. *Bull. Chem. Soc. Jpn.* 55: 2264-2266, 1982.
21. Moore, R.E. and Scheuer, P.J.: Palytoxin: A new marine toxin from a coelenterate. *Science* 172: 495-498, 1971.
22. Moore, R.E. and Bartolini, G.: Structure of palytoxin. *J. Am. Chem. Soc.* 103: 2491-2494, 1981.
23. Usami, M.; Satake, M.; Ishida, S.; Inoue, A.; Kan, Y. and Yasumoto, T.: Palytoxin analogs from the dinoflagellate *Ostreopsis siamensis*. *J. Am. Chem. Soc.* 117: 5389-5390, 1995.
24. Matsunaga, S., Fusetani, N., Hashimoto K., Koseki, K., Noma, M., Noguchi, H. and Sankawa, U.: Further kabiramides and halichondramides, cytotoxic macrolides embracing trisoxazole, from the *Hexabranchus* egg masses. *J. Org. Chem.* 54: 1360-1363, 1989.
25. Matsunaga, S., Fusetani, N., Hashimoto, K., Koseki, K. and Noma, M.: Kabiramide C, a novel antifungal macrolide from nudibranch egg masses. *J. Am. Chem. Soc.* 108: 247-849, 1986.

SCREENING METHODS FOR DETECTION AND EVALUATION OF BIOLOGICAL ACTIVITIES OF PLANT PREPARATIONS

A.J. VLIETINCK
Department of Pharmaceutical Sciences, University of Antwerp (UIA)
Universiteitsplein 1, B-2610 Antwerp, Belgium

Abstract

One of the important strategies and tactics in the search for new drugs still constists in the empirical approach i.e. the systematic screening of pure compounds or plant extracts in order to find useful lead compounds.

Structure-activity relationship studies of these lead compounds combined with computer-graphic model building can end up with promising candidate drugs having optimal activity and bioavailability, an acceptable therapeutic index and fewer side effects.

One of the key steps in the implementation of this rational serendipity methodology is the development of screening bioassays, which are adaptable to the testing of plant extracts, which can be highly colored, tarry, water insoluble and chemically very complex.

Several general screening bioassays including primary and broad bioassays and different specialized screening bioassays including lower organisms, isolated subcellular systems, isolated cellular systems, isolated organs of vertebrates and whole animals are systematically reviewed.

Introduction

Plants, especially medicinal plants, offer a vast resource of novel natural compounds, often with exciting activities and biological properties. Therefore, the emperical approach to discover new drugs *viz* the systematic screening of plant extracts or plant-derived substances still remains an interesting strategy to find new lead compounds. Structure-activity relationship studies of these leads preferentially combined with computer-graphic model building should result in molecules with optimal activity, better bioavailability, fewer side effects and an acceptable therapeutic index and consequently in good candidates for the development to new drugs.

Roles of bioassays

It is self-evident that in order to exploit plant material fully, an endeavour should be made to design as many screening programmes as possible. These screening programmes will of course be substantially different whether they are organized by a major pharmaceutical company, a consortium group consisting of several university research groups, all or not in combination with a research group of a large pharmaceutical company, or by a small pharmaceutical firm or an individual university research group. It is therefore necessary to put the bioassays to be discussed in the context of their usage before one can evaluate them or understand their benefits and weaknesses.

According to Suffness and Pezzuto [1] four major roles of bioassays can be distinguished i.e. prescreens, screens, monitors and secondary testings. In a prescreen a bioassay is applied to large numbers of initial samples to determine whether or not they have any bioactivity of the desired type. Such bioassays must have high capacity, low cost and must give rapid answers. They need not be quantitative. A bioassay in a screen is used to select materials for secondary testing, whereas in a monitor a bioassay is used to guide fractionation of a crude material towards isolation of the pure bioactive substances. It must, therefore, be fast and cheap, have high capacity and be readily available to the phytochemist. In the secondary testing, lead compounds are evaluated in multiple models and test conditions to select candidates for development towards clinical trials. Secondary testing is consequently characterized by a low capacity and expensive and slow bioassays.

Not all screening programmes contain separate prescreens and screens, since if the screening assay has enough capacity and is selective enough to reduce actives to a manageable number for the secondary testing to follow, prescreens are not needed. The monitor may be the same as the prescreen or the screen or may be a bench-top bioassay, which can be carried out using simple facilities in a chemistry laboratory.

Special considerations should be taken into account for screening and/or monitoring plant extracts. The methodology should be adaptable to highly colored, tarry, in water poorly soluble and chemically complex materials. Besides general requirements such as validity, lack of ambiguity, accuracy, reproducibility, simplicity and reasonabless of cost, the bioassays should be highly selective to limit the number of leads for secondary testing, highly specific to eliminate false positives and highly sensitive to detect also low concentrations of active compounds [2].

Most of the aforementioned requirements are better met by *in vitro* testing, so that in most screening programmes the typical pattern is that *in vitro* screens feed into *in vivo* tests, the latter mostly being introduced during the secondary evaluation. In all cases, however, a screening process must reduce the test substances for secondary evaluation to a manageable size, but hopefully not too few that no useful development candidates come out. The stage at which *in vivo* tests will be introduced largely depends on the hit rate of the screen. If there are a substantial number of actives, *in vivo* testing will be early whereas if actives are rare, they may go to final isolation and structure elucidation prior to *in vivo* testing [1].

The concept of a hit rate is, however, a difficult issue in many pharmaceutical companies. If a hit rate is set too high, a plethora of actives will result which will have to go through a second screening stage i.e. a successive combination of a subcellular and a cellular assay or vice-versa. If the hit rate is set too low, insufficient lead structures for isolation and structure-activity studies may be generated [3].

Classification of bioassays

The methods for the detection of biological activity of e.g. plant extracts can best be divided into two groups for screening purposes: general screening bioassays and specialized screening bioassays. Depending on the aims of the screening programme, either a general screening which can pick up many different effects or a specific assay which is directed at finding some effect against a specific disease has to be performed.

A broad screening bioassay is probably most useful if one is randomly screening chosen organisms for any kind of pharmacological activity. The alternative to using broad screening would be the setting up of a battery of specific test methods, which is cumbersome and expensive. Another drawback of using a general test for screening and monitoring is that one does not know, until the active compound has been isolated, if the work was worth doing. Since in most phytochemical laboratories engaged in the bioassay-guided isolation of actives from plant extracts no complex bioassays can be used, efforts have been made to introduce simple, inexpensive "front-line" or "bench-top" bioassays for the rapid screening of plant extracts and fractions. Care must be taken in the interpretation and predictive ability of these screening bioassays, but in general, they provide interesting preliminary information on the pharmacological potential of the plant extracts under study [4].

Whereas primary screening bioassays can be applied in-house by chemists, pharmacognosists, botanists and others, who lack the resources or expertise to carry out more elaborate bioassays, many specialized screening bioassays have been or are being developed; however, they are increasingly sophisticated and require the skills and expertise of biologic scientists.

Specialized screening bioassays can be subdivided according to the target organisms, which are used in the model [5]. These can be lower organisms e.g. microorganisms, insects, molluscs, protozoa, helminths, isolated subcellular systems e.g. enzymes, receptors, organelles, isolated intact cells of human or animal origin, isolated organs of vertebrates, or whole animals.

In any case, specialized bioassays have to be relevant that means they should predict the intended therapeutic indications. To be relevant or "correlational" a bioassay has to fulfill some basic criteria. First, the bioassay must be sensitive in a dose-dependent fashion to standard compounds that are known to possess the desired therapeutic property. Second, the relative potency of known active agents in the bioassay should be comparable to their relative potency in clinical use. Third, the bioassay should be selective i.e. the effects of known agents in this therapeutic indication should be distinguishable from effects of drugs for other indications [6].

It is clear that the degree of relevance increases from subcellular systems (molecular assays) to cellular systems (cellular assays), to organs up to conscious animals and human volunteers. Although considerable discussion is going on about the necessity of animal experiments, it is without any doubt that they are necessary for the discovery and evaluation of drugs. However, they should be performed only if they are necessary and well conceived [6,7].

Molecular bioassays, which use isolated subcellular systems have some contrasting characteristics with cellular bioassays, which use intact cells. The primary feature of molecular bioassays is their high specificity. It is possible to find a small number of new compounds with a specific activity in a selective way in a relatively short period of time using a high capacity receptor binding or enzyme inhibition assay. The hit rate will be quite low, however, and a large number of diverse samples need to be screened. All compounds working by a mechanism unrelated to the assay will be missed, whereas agents that do not enter the cells or are rapidly metabolised will be recorded as false positives. The cost of screening, however, is low but the cost of the large number of required samples must be considered. This type of screening is ideal in the first step in testing a specific hypothesis about the potential of agonists or antagonists of a particular molecular target to demonstrate pharmacological activity [1].

Examples of bioassays

1. GENERAL SCREENING BIOASSAYS

1.1. Broad screening bioassays

Broad assaying procedures can be performed by Hippocratic screening [8,9] or by the use of the isolated guinea-pig ileum [9, 10]. The Hippocratic method is performed on intact rats and involves observation of about 30 parameters at 3 to 5 dose levels. It is a time-consuming and expensive method, which requires much experience in observation and comparatively large amounts of test substances, but the screening of a wide variety of biological activities is possible. The large amount of extract and the time factor makes the method unsuitabe for monitoring the isolation of an active component from plant extracts. Substituting mice for rats and observation of only the most pronounced effect in a limited number of animals could potentially overcome the drawback of using the Hippocratic screening for monitoring purposes [9].

The isolated organ method, using e.g. guinea pig ileum for the observation of contraction or inhibition of contraction has been used as a non-specific screen for pharmacological effects of crude plant extracts by different research groups. Two tests are performed for each extract. First, the extracts ability to contract the ileum is tested. In a second test, the ileum is induced to contract by electrical stimulation and the extract is tested for ability to inhibit these contractions. As the ileum is a piece of smooth muscle, innervated by many different nerve systems, all of which can be electrically stimulated at the same time, a positive response can indicate many different mechanisms for the pharmacological activity of the extract. The most pronounced advantage of this test is that it is very simple and quick to perform and that it requires only small amounts of substance to be tested. The test is very appropiate for monitoring a subsequent procedure for isolation of pharmacologically active compounds. Although some information can be obtained by skilful application of known agonists and antagonists, the major drawback of being rather non-specific is still valid so that much time can be spent on isolation of compounds which turn out to be of limited pharmacological value [9].

1.2. Primary screening bioassays

The two most popular primary screening bioassays include the brine shrimp lethality test and the crown-gall tumour inhibition test [4,11]. The first technique is an *in vivo* lethality test in a tiny crustacean, the brine shrimp (*Artemia salina*). Since its introduction in 1982 [12], this test has been used for the isolation of *in vivo* active antitumour agents and pesticides produced by plants. It can, however, also been used to evaluate plants for different pharmacological activities, taking into account the basic premise that pharmacology is simply toxicology at a lower dose. Toxic substances might indeed elicit, at lower non-toxic dose, interesting pharmacological effects [11].

The second technique is the inhibition of the development of crown gall tumours on discs of potato tubers [13]. Crown-gall is a neoplastic disease induced by the gram-negative bacterium *Agrobacterium tumefaciens*, a bacterium known to infect a number of crop plants, and is due to the transfer of a tumour-inducing plasmid from the bacterium in the plant genome. The bioassay is fairly accurate in predicting 3PS (P388) *in vivo* murine antileukemic activity. It is not meant to replace the P388 assay, but it is a convenient, rapid, inexpensive, safe and statistically reliable prescreen for 3PS antitumour activity. The assay also circumvents mouse toxicity, which is an inherent disadvantage of the 3PS assay [11].

Another primary bioassay which is useful for determining which substance should be investigated for *in vivo* antineoplastic activity is the starfish or sea urchin assay. The eggs of the starfish, *Asterina pectinifera* and the sea urchin, *Strongylocentrotus purpuratus* have cell membranes permeable to a variety of substances. Exposure of the fertilised eggs to different antineoplastic agents will lead to different outcomes, which will allow to detect DNA and RNA synthesis inhibitors, microtubuli assembly inhibitors, and protein synthesis inhibitors. The sea urchin assay appears to be less selective than the starfish egg assay and relatively insensitive to a series of antineoplastic agents [4].

Other primary screening bioassays involve the detection of antibacterial and antifungal activities. There are a number of simple tests for antibiotic activity that can be carried out with simple equipment and a minimum of microbiologic expertise. In agar dilution streak assays, up to seven different organisms can be screened simultaneously on a petri dish at a fixed concentration of extract. Weak antimicrobial agents present in low concentrations (<1%) can be detected. The introduction of the microtitre plate, containing e.g. 96

holes, allows the simultaneous testing of three dilutions of one or two extracts against not less than 24 or 12 microorganisms, respectively [2].

Representative microorganisms responsible for human infections of significance can be chosen for screening. A prototype of a microbial battery for screening plant extracts consists of one or more representatives of the different groups of bacteria including gram-positive and gram-negative cocci and rods and one or more representatives of yeasts, spore-producing fungi and dermatophytes.

Some simple and quick bioassays monitor the inhibitory effects of extracts on seed germination. The etiolated wheat coleoptile bioassay has been used not only to detect plant growth regulating substances but also to find compounds that have mycotoxic, immuno-suppressant and antifungal activity [14]. It is important, however, to be aware of the fact that some compounds that are inhibitory at high concentrations can be growth promoting at lower concentrations, and vice versa.

Also herbicidal-, insect-antifeedant, larvicidal- and molluscicidal activities can been determined by simple bioassays, which can function as surrogate assays to isolate bioactive compounds from plant extracts. It should, however, be reminded that primary bioassays provide only preliminary information, which always should be checked in more appropiate specialized bioassays.

2. SPECIALIZED SCREENING BIOASSAYS

As described above, after having found a certain type of activity, it will be necessary to study this activity in more detail by using one or several specailized bioassays for screening and/or monitoring purposes. These *in vitro* or *in vivo* tests are more sophisticated than the primary screening bioassays and require the expertise of biochemists or pharmacologists. They can consequently only been performed in a multidisciplinary team.

We have classified the many existing specialized screening bioassays according to the target organisms, which are used. Some of these bioassays are described in this section.

2.1. Lower organisms

Numerous research programmes to detect and isolate antibacterial-, antifungal-, antiviral- and antiparasitic compounds from plants and other natural sources have been designed and are performed. Large batteries of bacteria, yeasts, fungi, viruses, insects, mollluscs, protozoa and helminths are thereby used in a broad range of *in vitro* and/or *in vivo* bioassays.

Relatively few antiviral drugs are avilable, since they must need a number of criteria. They must inhibit at least one of the propagation steps of the virus, have a broad range of activity 'and not be immunosuppressive. The methods commonly used for evaluation of *in vitro* antiviral activities are based on the different abilities of viruses to replicate in cultured cells. Some viruses can cause cytopathic effects (CPE) or form plaques. Others are capable of producing specialized functions or cell transformation. Virus replication in cell cultures may also be monitored by the detection of viral products i.e. viral DNA, RNA or polypeptides. Thus, the antiviral test selected may be based on inhibition of CPE, reduction or inhibition of plaque formation and reducing virus yield or other viral functions [15]. It should be emphasized that the toxic effects of the antiviral agent on the host cells must be considered since a substance may exhibit an apparent antiviral activity by virtue of its toxic effects on the cells. The cytotoxicity assay on cell cultures is usually done by the cell viability assay and the cell growth rate test, although other parameters such as destruction of cell morphology under microscopic examination or measurement of cellular DNA synthesis have been used as indicators of compound toxicity [16].

The U.S. National Cancer Institute (NCI) has in 1987 implemented AIDS-screening and antiviral development programmes in parallel with its long standing anticancer drug programme. The effects of HIV and potential antiviral agents on human CEM T4-lymphocytes are followed by cell survival, which is determined by means of the vital stain XTT (2,3-bis[1-methoxy-4-nitro- 5-sulfonylphenyl]- 5-[(phenylamino) carbonyl]-2H-tetrazoliumhydroxide), a soluble version of the tetrazolium salt MTT. Dose-response data are produced automatically from the optical densities obtained on the plates in an automatic plate reader using especially developed software. The drug zidovudine (AZT) is used as a positive control for quality control purposes and as standard for comparisons [17]. Increasing attention is being focused on inhibitors of several HIV-enzymes such as reverse transcriptase, integrase and protease [18].

Animal models for a number of virus infections are available. They are helpful in detecting not only if the candidate compound is an effective viral inhibitor without inducing viral resistance but also if it has a good bioavailability, resists and not adversely effects the immune system and not interferes with the normal metabolic processes of uninfected cells [14]. Some viral infections, such as herpes simplex in guinea-pig and mice and cytomegalo- or poliovirus infections in mice, mimic the natural diseases very closely. These animal models allow continuous observation of the infected animals, and virus spread and pathogenesis at local sites can be studied thoroughly [20]

Protozoal diseases are a major threat to world health and they included malaria, leishmaniasis, trypanosomiasis, amoebiasis and giardiasis. In addition, cryptosporidosis, pneumocystis and toxoplasmosis are becoming more prevalent in developed countries due to suppression of the immune system.

Two major antimalarial screening tests using the rodent parasite *Plasmodium berghei* in mice may be used for the screening of plant extracts, the 4-day suppressive test of blood schizontocidal action (Peter's test) and the Rane test of blood schizontocidal activity.

Neither of these tests is suitable for the identification of long acting compounds. Useful *in vitro* tests have become available only recently. The inhibition of the incorporation of ^{3}H-hypoxanthine into plasmodia of the human malaria parasite, *Plasmodium falciparum*, can be quantitatively measured in microplates and has been adapted to a semi-automated procedure. The data obtained with this method have been shown to correlate with *in vivo* data obtained with *P.berghei* [21].

Since *in vivo* testing for amoebiasis is notoriously difficult, unpleasant to perform, expensive and time consuming, several *in vitro* models have been developed. In their initial tests with *Entamoeba histolytica* Philippson et al. [22] used flat sided culture tubes and assessment of activity was obtained by counting the amoebae. Subsequently, a more sensitive microplate was developed and amoebal growth was measured by staining with eosin. Similarly, a new microplate assay was developed for determining antigiardial activity *in vitro* utilizing *Giardia intestinalis* and measuring soluble formazon production from a tetrazolium reagent.

Although it is not possible to develop antimalarial and amoebicidal drugs without resort to *in vivo* procedures, the *in vitro* tests which are now available can be used to select new drug leads from plants. By combining *in vitro* cytotoxicity testing with *in vitro* amoebicidal and antiplasmodial testing it is possible to obtain some measure of selectivity of action [21].

Schistosomiasis, known as bilharzia, is a parasitic disease, propagated by three species of schistosomes *viz*. *Schisotosoma mansoni* and *S.japonicum* (intestinal bilharzia) and *S.haematobium* (urogenital bilharzia). The transmission of schistosomiasis requires the presence of an intermediate host such as a fresh water snail of the genera *Biomphalaria* and *Bulinus*. Plant extracts can be investigated for their toxicity to *Schistosoma mansoni* in mice. Schistosomes are isolated from mice by a simple perfusion technique and collected after sedimentation in saline. For counting, adult worms can be recovered on a mesh screen. The livers of the mice are examined for dead worms and the eggs in the intestines and livers are counted. The overall procedure lasts at least 12 weeks and is obviously not suitable for bioassay-guided isolations [23].

2.2. Isolated subcellular systems

Molecular assays look for activity using isolated systems including enzymes and receptors. Through already known for many years in pharmacology and frequently used in industrial drug development, in studies of natural products the systematical use of these methods for screening purposes has only started.

Only a limited number of the many existing enzyme assays have been utilized in the screening of extracts [6]. Examples of such biochemical test systems include macrophage associated carboxypeptidases such as trypsin and kallikrein, enzymes of the arachidonic pathway such as cyclooxygenase, 5-lipoxygenase and thromboxan-synthetase and enzymes of the hypertensive proteolytic cascade such as renin and angiotensin converting enzyme (ACE) and the genital proteolytic system such as acrosin.

Other enzyme assays, which have been used for the screening of natural substances include monoamine oxidase (antidepressant activity), H^+/K^+-ATP-ase (anti-gastric ulcer activity), polysaccharide degrading enzymes (amylase, glucosidase) and aldose reductase (anti-diabetic activity) HMG-CoA reductase (anti-atherosclerotic activity), 5α-reductase (anti-androgenic activity), and xanthine oxidase (anti-oxydative activity).

The inhibitory activity of the test compounds is mostly determined after chromatographic separation of the enzymatic reaction products by means of UV or fluorescence spectrophotometric detection.

Many drug discovery programmes have been based on the utilization of radioligand receptor binding technology to identify lead compounds which interact with receptors likely to be important in neuronal, immunological, gastrointestinal and cardiovascular function/ dysfunction. The use of radio-isotope-labeled transmitters or drugs (radioligands) resulted in the development of a simple and sensitive methodology, which provides information detailing the molecular interactions between drugs and receptors. Radioligand binding assays are reliable, simple, rapid and very sensitive, which make them very suitable as bioassays for screening and monitoring of plant extracts. Typically, a specific binding of >75% is needed for high volume screening and should accurately reflect the interaction between the drug and the receptor, and not be a non-specific binding interaction, for it is important that the assay be run at equilibrium. Sweetnam et al. [14] have described the role of receptor binding in drug discovery and illustrated this by presenting the comparative data from 38 different primary assays including receptors for adenosine, catecholamines, amino acids, biogenic amines, channel proteins, acetylcholine, opiates, prostanoids, reuptake sites and second messengers.

Two screening studies of plant extracts, used in traditional medicine of China and Suriname respectively, for receptor binding activity have recently been published [25, 26]. In both studies the results provide partial scientific support for a number of plant species for their use in traditional medicine for the relief of pain [25], against proteinuria [26,27] and for their tranquillizing effects [28]. Bioassay-guided fractionation using ligand binding studies resulted in the isolation of adenosine-A1-active and 5-HT_{1A} active ligands [27, 28].

Although receptor ligand binding assays are sensitive and specific they are not necessarily predictive of activity *in vivo* and any positive findings would require further verification by the use of functional assays in whole animals or with isolated organs.

Very recently, the on-line coupling of reversed-phase liquid chromatography to a biochemical detection system based on receptor ligand interactions has been realized. The receptor-affinity detection (RAD) is performed using a post column reaction detection system with open-tubular reaction coils. When combined with structure-activity elucidating methods like mass spectrometry (MS) the RAD-system should be able to give two dimensional spectra on structure and activity. Using an combined LC-RAD/MS method, the biochemical activity and structure elucidation can be performed in one run. It will, however, be a great challenge to implement other biological affinity interactions in LC-RAD systems e.g. membrane bound receptors and cell-bound receptors, since up till now only soluble receptors have been tested in the post-column detection system [29, 30].

2.3. Isolated cellular systems

Cellular assays utilizing intact cells of human or animal origin have been used for more than half a century to screen for cytotoxic agents from natural sources. Cytotoxicity, or toxicity to cells in culture, can be subdivided into cytostatic activity i.e. stopping cell growth (often reversible) and cytocidal activity i.e. killing cells. A large number of cytotoxicity-based bioassays have been used as prescreens for antitumour and antineoplastic activity. These short-term *in vitro* growth inhibition assays with cultured cells include Walker carcinosarcoma 256, mouse L-120 leukemia, Ehrlich murine ascites tumour, sarcoma 180 and mouse P-388 leukemia cell lines [31]. While these traditional cell-cytotoxicity assays are indicative for activity in leukemia, lymphoma, or a few rare tumours, their efficacy in finding products with activity in the predominantly occurring slow growing solid tumours of humans is strictly limited. Therefore, the US NCI has in 1989 developed a disease-oriented strategy in which an *in vitro* primary screen employing a tumour cell line panel consisting of a total of 60 known tumour cell lines, derived from seven cancer types i.e. lung, colon, melanoma, renal, ovaria, brain and leukemia, is used [32]. The objective is the discovery of agents that demonstrate selective cytotoxicity with a cell line derived from a single type of primary human tumour. This *in vitro* disease-oriented prescreen can be followed by an *in vivo* screen, since the cells can be carried out as solid tumours in athymic mice [1]. Some practical considerations and applications of this *in vitro* anticancer drug discovery screen have recently been reviewed [33].

Several mechanism-based *in vitro* bioassays can be designed by analogy with the types of molecular responses mediated by known, clinically effective antitumour agents. Thus, as examples, monitoring effects similar to those known to be mediated by vinblastine (tubulin depolymerization), taxol (tubulin stabilisation), camptothecin (topoisomerase inhibition), 2-methyl-9-hydroxy-ellipticinium (topoisomerase II inhibition) and bleomycin (DNA cleavage) are all reasonable avenues toward novel drug discovery. To establish the true efficacy of promising anticancer agents, subsequent evaluation in more advanced testing systems is required, followed by (pre)clinical trials [1].

Chemoprevention of cancer may be defined as the deliberate introduction of selected substances into the diet for the purpose of reducing cancer incidence. A large number of plant metabolites may, indeed, have cancer-preventive properties [34]. It can thereby be anticipated that anticarcinogens act on any of the transition states between exposure to a carcinogenic agent and the final outcome (disease: cancer). According to Wattenberg [35] cancer-preventing substances can be classified into three categories according to the phase time in carcinogenesis when they are effective *viz.* compounds that prevent the formation of carcinogens from precursor substances, compounds that prevent carcinogenic agents from seaking or reacting with critical target sites in tissues (blocking agents) and agents that act subsequent to exposure to carcinogens, once the initiating damage has been irreversibly done (suppressing agents). With respect to the onset of carcinogenesis, initiation, Hartman and Shankel [36] describe various levels at which antimutagenesis is effected like desmutagenesis (destruction/inactivation of mutagens), interference with error-prone and stimulation of DNA-repair, and several other possibilities

The establishment of a number of short-term tests for genotoxicity e.g. chromosome aberration test, micronucleus test has provided the possibility to predict whether or not a substance possesses an intrinsic mutagenicity and thus the potential to affect the initiation phase of carcinogenesis. These assays are then valuable to select potential mutagens. Conversily, these assays can also be used to get an impression of the possible antimutagenic potential of a compound by assessing its modulatory effects on the results of established mutagens in these assays : a decrease in the response of these positive controls may point towards potential antimutagenicity. The tests can be conducted either *in vitro* with prokaryotes or eukaryotes cell systems or *in vivo* with experimental animals. Based upon *in vitro* and *in vivo* data activity profiles of antimutagens can be developed [37]. The most valuable data on cancer-preventing effects of plant products may come from studies on humans. They can be performed in two ways *viz.* epidemiology studies based on dietary questionnaires and experimental biomarker research, which links toxicologal and epidemiological studies [34].

Several nutritive dietary anticarcinogens include vitamins A, C, E and β-carotene, all well-known antioxidants and/or radical scavengers, which could serve as an explanation for this mode of action. Many experiments have indeed shown that reactive oxygen species (ROS) including radicals and oxidants play various roles in the development of cancer. ROS can directly damage DNA, activate transcription factors, or active kinases and each of these changes may activate various genes including oncogenes. Gene activation may also be mediated by cytokines released in response to ROS or from phagocytic cells [38].

There is compelling evidence that oxidative stress is implicated in the pathogenesis of many disorders besides cancer, although a consensus for such involvement has been reached only in a few cases. A role for free radicals has been postulated for some cardiovascular diseases such as atherosclerosis and reperfusion injury, several neurological disorders such as Alzheimers' and Parkinsons' diseases, some lung diseases such as fibrosis and emphysema, cataract, rheumatoid arthritis and other inflammatory diseases and also aging.

The finding of new antioxidative leads which are capable of preventing and/or eliminating oxidative stress and development of suitable screening bioassays merits consequently high priority and would be very welcomed.

An survey on *in vitro* and *in vivo* assays for ROS has been published by Halliwell et al. [39]. These techniques include trappic assays, in which the radicals are allowed to react with a trap molecule to give one or more stable products, which are then measured, or with a spin trap to form a more-stable radical, which is detectable by electron spin resonance (ESR).

Other useful screening bioassays are fingerprint assays including the "DNA fingerprinting" approach, in which the DNA damage that occurs when cells and tissues are subjected to oxidative stress is chemically determined, and the measurement of the end products of lipid peroxidation such as lipid hydroperoxides and certain aldehydes. Measurements of some of the end products of oxidative damage to proteins can be achieved by the protein carbonyl assay. GC, HPLC, HPLC/MS, HPLC/antibody technique, light emission, fluorescence and several enzymatic reactions are the methods used to detect and measure biological lipid peroxidation [39].

Another important field for which isolated cellular systems are used in many screening bioassays is that of immunostimulation. It has to be emphasized that immunostimulation primarily implies stimulation of the non-specific immune system i.e. stimulation of the functions and thus the efficiency of granulocytes, macrophages, Kupfer cells, monocytes, natural killer cells, complement factors and certain T-lymphocyte populations. This also includes the stimulation of a series of mediators such as mono- and lymphokines secreted by these cells in response to any induction [40,41]. Since many recurrent infections and malignant diseases are caused by a decreased number of immune competent cells, it is clear why granulocytes , monocytes, macrophages and T-lymphocytes are the preferred target cells for screening. These cells can be obtained from human donor blood or animal organs. A survey of assays for immunomodulation is described by Wagner and Jurcic [41].

In vitro tests such as the granulocyte phagocytosis assay (smear test), the chemoluminescence assay, the chemotaxis assay, the lymphocyte proliferation assay, the assay of natural killer activity and the assay for tumour necrosis factor (TNP) production as well as *in vivo* tests such as the carbon clearance assay for phagocytosis have extensively been utilised in the screening and monitoring of active plant compounds. The *in vivo* assay for anticomplement activity has not only a relevance for the processing of antigens but also for inflammatory processes and can be used as a screening or monitoring for anti-inflammatory active plant extracts [42, 43].

Other *in vitro* assays which have been utilized by our research group to screen and monitor active plant extracts include the inhibition assay of platelet aggregation as experimental model for evaluating the antithrombotic and anti-inflammatory potential of the isolated compounds [44, 45] and the proliferation assay of human umbilical vein endothelial cells (HUVEC) as experimental *in vitro* model for wound healing [46, 47].

2.4. Isolated organs of vertebrates

Organ based assays are more and more replaced nowadays as the front-line primary screen, but often retain signification as secondary screens to confirm the results of e.g. radioligand-binding bioassays and hence to assist in prioritization of active extracts or compounds. They represent consequently an essential connection between the high technology of the primary screens and the realities of the pharmacological effectiveness [3]. Segments of the gastro-intestinal tract or spirally cut strips of vascular tissue are mainly used in the organ bath method for isolated organs. Since the guinea-pig ileum gives contractions to many agonists including acetylcholine, angiotensin, arachidonic acid, bradykinin, histamine (H_1-receptors), prostaglandins (type E), serotonin and substance P, this isolated organ has been used as a broad screening bioassay for non-specific antispasmodic activity and for parasympathicomimetic or parasympathicolytic activity (see under 1.1). For the study of adrenergic mechanisms the isolated rabbit jejunum is appropiate, whereas the rat phrenic nerve-diaphragm preparation is utilized for the study of the action of muscle relaxants [7]. With all these models single-dose experiments, cumulative dose-response experiments and coaxial stimulation experiments can be carried out [48, 49].

The tracheal spiral from sensitised guinea-pigs is often used to represent the large airways; lung parenchyma strips, prepared from the same animals are then utilized to represent peripheral airways. Contractions can be induced with histamine, carbachol, LTD_4, PAF and arachidonic acid, and the effects of plant extracts or tested against these directly [7].

Possible vasodilating effects can be studied using the rabbit central ear artery perfusion model [50]. In the rat mesenteric artery model, the perfusion system is connected to a suitable device for detecting changes in perfusion pressure to indicate vasoconstriction or vasodilation [7]. Cardiotonic properties can be studied by using the spontaneously beating right atrium or the stimulated left atrium of the guinea-pig [50]. The superfusion technique, which consists of bathing an isolated tissue with a stream of artificial salt solution at 37°C and a constant flow rate allows the testing of several tissues in cascade, generally up to six arranged in two banks. This arrangement allows a parallel assay of individually injected samples. We have used such an experimental set-up for the detection of prostaglandin-like activities of plant extracts. The cascade consisted of two vascular smooth muscle preparations *viz.* the rabbit coeliac and mesenteric arteries and one non-vascular smooth muscle preparation *viz.* the rat stomach strip. Prostaglandin-E like activity was detected as vascular smooth muscle relaxing and non-vascular smooth muscle stimulating activities in the cascade model [51].

A cascade system consisting of rabbit coeliac and mesenteric arteries and rabbit aortic tissue was used for the detection of serotoninergic activity in plant extracts. The system can even be refined by fixing two separate aortic ring preparations, one with denuded endothelium and the other with intact endothelium on the cascade. In this way, relaxation due to the release of endothelium-derived relaxing factor (EDRF) or nitric oxide can be determined [52].

2.5. Whole animals

Biomedical research needs animals. This is most obvious in case of *in vivo* animal experiments. However, for other scientific reasons e.g. *in vitro* studies, biological material is also needed to study enzymes, membranes, receptors, cells, tissues or organs which are obtained from dead animals. Therefore, animals have to be sacrificed in biomedical laboratories [i] at the end of *in vivo* experiment (ii) during experiments where sacrifice of the animals is not part of the study but must be done when pain, distress and suffering exceed acceptable levels or if it is likely for the animal to remain in pain or distress after cessation of the experiment, and [iii] to provide biological material for *in vitro* studies [6].

Most large pharmaceutical companies dispose of a large battery of *in vivo* tests for the secondary evaluation of selected leads. Animals used in these tests include mice, rats, hamsters, guinea-pigs, rabbits, cats, dogs, ferrets, cattle, sheep, goats, horses, pigs and primates.

In vivo tests for the detection of activity of the cardiovascular system, the gastrointestinal tract, the liver and the biliary system, the respiratory system, the renal system, and the endocrinological system as well as for the determination of anti-inflammatory, analgesic, psychotropic, neurotropic, immunomodulating and antidiabetic activity and effects on learning and memory have recently been reviewed [6]. *In vivo* tests also remain in many cases the stepping-stone between antimicrobial, antiviral, antiparasitic and anticancer *in vitro* tests and the demonstration of a corresponding activity in human clinical trials. These models should predict efficacy in man, and must therefore mimic the natural disease as closely as possible. The challenge for the investigator always will be to correlate *in vitro* data with *in vivo* findings, bearing in mind the old saying "*In vitro* simplicitas, *in vivo* veritas".

Conclusions

An attempt has been made to outline the most important aspects of the empirical approach to find new lead compounds from plants. It is thereby apparent and promising to state how much progress has been made in the development of sensitive and simple primary and specialized bioassays for screening and/or monitoring purposes of plant extracts. The pharmaceutical industry has now the disposal of automated high-throughput *in vitro* screens for biological activities which are capable of examining thousands of compounds or extracts in very short periods of time.

As a result of the new technologies, it has become current fashion to guide fractionation of plant extracts towards isolation of the pure bioactive compounds. The expectation that in the near future several new lead chemical entities of natural origin will be placed in the research and development pipeline, is therefore not too presumptuous.

References

1. Suffness, M. and Pezzuto, J.M. (1991) Assays related to cancer drug discovery in K. Hostettmann (ed.), *Methods in Plant Biochemistry,* Vol. 6, *Assays for Bioactivity*, Academic Press, London, 71-153.
2. Vanden Berghe, D.A. and Vlietinck, A.J. (1991) Screening for antibacterial and antiviral agents in K.Hostettmann (ed.), *Methods in Plant Biochemistry*, Vol. 6, *Assays for Bioactivity,* Academic Press, London, 47-69.
3. Cordell, G.A. (1995) Changing strategies in natural products chemistry, *Phytochemistry,* **40**, 1585-1612
4. Ghisalberti, E.L. (1993) Detection and isolation of bioactive natural products in S.M. Colegate and R.J. Molyneux (eds.), *Bioactive natural products: detection, isolation and structural determination*, CRC Press, Boca Raton, 9-57.
5. Hostettmann, K., Marston, A. and Wolfender, J.L. (1995) in K.Hostettmann, A. Marston, M. Maillard and M. Hamburger (eds.), Proceedings of the Phytochemical Society of Europe, Phytochemistry of Plants used in Traditional Medicine, Clarendon Press, Oxford, 17-45.
6. Vogel, H.G. and Vogel, W.H. (eds. (1997) Drug discovery and evaluation: pharmacological assays, Springer Verlag, Berlin, 1-755.
7. Williamson, G.M., Okpako, D.T. and Evans, F.J. (1996) in Pharmacological Methods in Phytotherapy Research, Vol. I : Selection, Preparation and Pharmacological Evaluation of Plant Material, John Wiley & Sons, Chichester, 1-228.
8. Malone, M.K. (1983) The pharmacological evaluation of natural products: general and specific approaches to screening ethnopharmaceuticals, *J. Ethnopharmacol.* **8**, 127-147.
9. Samuelsson, G. (1991) Assays for pharmacological activity: non-specific assays in K.Hostettmann (ed.), *Methods in Plant Biochemistry*, Vol. 6, *Assays for Bioactivity*, Academic Press, London, 261-280.
10. Vlietinck, A.J., Vanden Berghe, D.A. and Laekeman, G.M. (1991) in Proceedings of the International Symposium on Innovations in Pharmaceutical Sciences and Technology, Shri B.V. Patel Perd Centre, Ahmedabad, 27.10-29.10.1990, Vol. I, 150-182.
11. McLaughlin, J.C. (1991) Crown gall tumours on potato discs and brine shrimp lethality: two simple bioassays for higher plant screening in K.Hostettmann (ed.), *Methods in Biochemistry*, Vol. 6, *Assays for Bioactivity*, Academic Press, London, 1-32
12. Meyer, B.N., Ferrigni, N.R., Putmann, J.E., Jacobson, L.B., Nichols, D.E. and McLaughlin, J.L. (1982) Brine shrimp: a convenient general bioassay for active plant constituents, *Planta Med.* **45**, 31-34.
13. Ferrigni, N.R., Putman, J.E., Anderson, B., Jacobson, L.B., Nichols, D.E., Moore, D.S., McLaughlin, J.L., Powell, R.G. and Smith, C.R. Jr. (1982) Modification and evaluation of the potato disc assay and antitumour screening of Euphorbiacaea seeds, *J. Nat. Prods.*, **45**, 679-686.
14. Kubo, L. (1989) Effects of a marine algal constituent on the growth of lettuce and rice seedlings, *Pure Appl. Chem.* **61**, 373-376.

15. Vlietinck, A.J. and Vanden Berghe, D.A. (1991) Can ethnopharmacology contribute to the development of antiviral drugs? *J. Ethnopharmacol.*, **32**, 141-153.
16. Hu, J.M. and Hsiung, C.D. (1989) Evaluation of new antiviral agents. I. *In vitro* perspectives, *Antiviral Res.* **11**, 217-232.
17. Lednicer, D. and Narayanan, V.L. (1993) Acquisition and screening of natural products as potential anticancer andantiviral agents in S.M.Colegate and R.J.Molyneux (eds.) Bioactive natural products: detection, isolation and structural determination, CRC Press, Boca Raton, 159-172.
18. De Clerq, E. (1995) Antiviral therapy for human immunodeficiency virus infections, *Clin. Microb. Rev.*, **8**, 200-239
19. Galasso, G.J. (1989) Antiviral agents. Why not a "penicillin" for usual infections? In E. De Clercq and R.T. Walker (eds.) Targets for the design of antiviral agents. Plenum Publishing Cooperation, New York, ...
20. Sidwell, R.W. (1986) Determination of antiviral activity, *Drugs and Pharmaceutical Sciences*, **27**, 433-480.
21. Philippson, J.D. (1991) Assays for antimalarial and amoebicidal activities in K.Hostettmann (ed.). *Methods in Plant Biochemistry*, Vol. 6 Assays for Bioactivity, Academic Press, London, 135-152.
22. Philippson, J.D. (1995) A matter of some sensitivity, *Phytochemistry*, **38**, 1319-1343.
23. Baldé, A.M., Van Marck, E.A., Kestens, L., Gigase, P.L. and Vlietinck, A.J. (1989) Schistosomicidal effects of *Pavetta owariensis* and *Harrisonia abyssinica* in mice infected with *Schistosoma mansoni*, *Planta Med.* **55**, 41-43.
24. Sweetnam, P.M., Caldwell, L., Lancaster, J., Bauer, L. Jr., McMillan, B., Kinner, W.J. and Price, C.H. (1993), The role of receptor binding in drug discovery, *J. Nat. Prods.*, **56**, 441-455.
25. Zhu, M., Greengrass, P.M. and Philippson, J.D. (1996) Application of radioligand receptor binding assays in the search for CNS active principles from Chinese medicinal plants, *J. Ethnopharmacol.*, **54**, 153-164.
26. Hasrat, J.A., De Backer, J.-P., Vauquelin, G. and Vlietinck, A.J. (1997) Medicinal plants in Suriname : screening of plant extracts for receptorbinding activity, *Phytomedicine*, 4, 59-65.
27. Hasrat, J.A., Pieters, L., Claeys, M. and Vlietinck, A. (1997) Adenosine-1 active ligands: cirsimarin, a flavone glycoside from *Microtea debilis*, *J. Nat. Prods.*, **60**, 638-641.
28. Hasrat, J.A., Pieters, L., De Backer, J.-P., Vauquelin, G. and Vlietinck, A. (1997) Screening of medicinal plants from Suriname for 5-HT_{1A} ligands: bioactive isoquinoline alkaloids from the fruits of *Annona muricat*, *Phytomedicine*, **4**, 133-140.
29. Irth, H., Oosterkamp, A.J., Tjaden, U.R. and van der Greef, J. (1995) Strategies for on-line coupling of immunoassays to high-performance liquid chromatography, *Trends Anal. Chem.*, **14**, 355-361.
30. Oosterkamp, A.J., Villaverde Herraiz, M.T., Irth, H., Tjaden, U.R. and van der Greef, J. (1996) Reversed-phase liquid chromatography coupled on-line to receptor-affinity detection based on the human estrogen receptor, *Anal. Chem.*, **68**, 1201.
31. Geron, G.T., Greenberg, M.M., McDonald, A.M., Schumacher, A.M. and Abbott, B.J. (1972) Protocols for screening chemical agents and natural products against animal tumours and other biological systems, Cancer Chemother. Rep. (Part 3), **3**, 1-14.
32. Boyd, M.R. (1989) Status of the NCI preclinical antitumor drug discovery screen. Principles and Practice of Oncology, PPO Updates, Vol. 3 (10), 1-12.
33. Boyd, M.R. and Paull, K.D. (1995) Some practical considerations and applications of the National Cancer Institute in vitro*ff* anticancer drug discovery screen, *Drug Developm. Res.*, **34**, 91-109.
34. Verhagen, H. and Feron, V.J. (1994) Cancer prevention by natural food constituents. The lessons of toxicology transposed to antigenotoxicity and anticarcinogenicity in H. Kozlowska, J. Fornal and Z. Zdunézyck (eds.) *Bioactive substances in food of plant origin*, **2**, 463-478.
35. Wattenbergh, L.W. (1992) Inhibition of carcinogenesis by minor dietary constituents, *Cancer Res. (suppl.)*, **52**, 2085s-2091s.
36. Hartman, P.E. and Shankel, D.M. (1990) Antimutagens and anticarcinogens: a survey of putative interceptor moles, *Environm.. Molec. Mutagen.*, **15**, 145-182.
37. Waters, M.D., Stack, H.F., Jackson, M.A., Brockman, U.E., De Flora, S. (1996) Activity profiles of antimutagens: *in vitro* and *in vivo* data, *Mutat. Res.*, **350**, 109-129.
38. Kehrer, J.P. and Smith, C.V. (1994) Free radicals in biology: sources, reactivities and roles in the ethiology

of human diseases, in B. Frei (ed)., Natural oxidants in human health and diseases, Academic Press, New York, **2**, 25-62.

39. Halliwell, B., Gutteridge, J.M.C. and Cross, C.E. (1992) Free radicals, antioxidants and human disease : where are we now? *J. Lab. Clin. Med.*, 598-620.
40. Wagner, H. and Jurcic, K. (1991) Assays for immunomodulation and effects in mediators of inflammation, in K. Hostettmann (ed.), Methods in Plant Biochemistry, Vol. 6, Assays for Bioactivity, Academic Press, London, 195-217.
41. Labadie, R.P. (1993) Immunomodulatory compounds, in S.M.Colegate and R.J. Molyneux, Bioactive natural products: detection, isolation and structure determination, CRC Press, Boca Raton, **14**, 279-317.
42. Lasure, A., Van Poel, B., Cimanga, K., Pieters, L., Vanden Berghe, D. and Vlietinck, A.J. (1994) Modulation of the complement system by flavonoids, *Pharm. Pharmacol. Lett.*, **4**, 32-35.
43. Cimanga, K., De Bruyne, T., Lasure, A., Van Poel, B., Pieters, L., Vanden Berghe, D. and Vlietinck, A. (1995) *In vitro* anticomplementary activity of constituents from *Morinda morindoides*, *J. Nat. Prod.*, **58**,372-378.
44. Rasheed, A., Laekeman, G.M., Vlietinck, A.J., Janssen, J., Hatfield, G., Tott¿)¿, J. and Herman, A.G. (1984) Pharmacological influence of nutmeg and nutmeg constituents on rabbit platelet-function, *Planta Med.*, **50**, 222-226.
45. Laekeman, G.M., De Clerck, F., Vlietinck, A.J. and Herman, A.G. (1985) Vernolepin: an anti-platelet compound of natural origin, *Naunyn-Schiemdeberg's Arch. Pharmacol.*, **331**, 108-113.
46. Vanden Berghe, D.A., Yang, Q.R., Totté, J. and Vlietinck, A.J. (1993) Specific stimulation of human endothelial cells by *Triticum vulgare* extract and its biologically active fraction, *Phytother. Res.*, **7**, 172-178.
47. Pieters, L., De Bruyne, T., Claeys, M., Vlietinck, A., Calomme, M. and Vanden Berghe, D. (1993) Isolation of a dihydrobenzofuran lignan from South American Dragon's blood (*Croton* spp.) as an inhibitor of cell proliferation, *J. Nat. Prod.*, **56**, 899-906.
48. Laekeman, G.M., Mertens, J., Totté, J., Bult, H., Vlietinck, A.J. and Herman, A.G. (1983) Isolation and pharmacological characterization of vernolepin, *J. Nat. Prodff.*, **46**, 161-169.
49. Bakana, P., Laekeman, G.M., Totté, J., Herman, A.G. and Vlietinck, A.J. (1985) Stereochemical considerations in relation to the pharmacological activity of *Pterotabernaff alkaloids*, *J. Nat. Prod.*, **48**, 766-771.
50. Laekeman, G.M., Claeys, M., Rwangabo, P.C., Herman, A.G. and Vlietinck, A.J. (1986) Cardiovascular effects of 3-methylquercetin, *Planta Med.*, **no. 6**, 433-437.
51. Üstünes, L., Claeys, M., Laekeman, G., Herman, A.G., Vlietinck, A.J. and Özer, A. (1985) Isolation and identification of two isomeric trihydroxy octadecenoic acids with prostaglandin E-like activity from onion bulbs *(Allium cepa)*, *Prostaglandins*, **29**, 847-865.
52. Üstünes, L., Özer, A., Laekeman, G.M., Corthout, J., Pieters, L.A.C., Baeten, W., Herman, A.G., Claeys, M. and Vlietinck, A.J. (1991) Clinical characterization and pharmacological activity of nazlinin, a novel indole alkaloid from *Nitraria schoberi*, *J. Nat. Prod.*, **54**, 959-966.

THE USE OF RECEPTOR BINDING, A VERY SPECIFIC, HIGH CAPACITY SCREENING METHOD, IN THE IDENTIFICATION OF BIOLOGICALLY ACTIVE COMPONENTS FROM NATURAL SOURCES

MARTIN TH. M. TULP
Solvay Duphar B.V., Department of Pharmacology,
PO Box 900, 1380 DA Weesp, The Netherlands

Nature has been a rich source of valuable drugs despite the fact that the exploration of natural sources has been largely guided by toxicology and ethnopharmacology. Most traditional medicines have been found by coincidence, or by trial and error, not by systematic research. Thus it is very likely that the majority of useful drugs in nature so far have escaped discovery. This, together with the fact that only a minute fraction of natural sources has been partially explored, leads to the conclusion that natural product research has barely begun. In vitro mechanistic screening techniques such as receptor binding, are very selective and sensitive analytical methods. In recent years, they have been highly automated and converted to high throughput screening procedures making them extremely powerful tools in drug research. Recent developments in synthetic organic chemistry, collected under the broad term 'combinatorial chemistry' result in an unprecedented output of new compounds which are subjected to high throughput screening procedures. Thus, natural product research is obliged to make use of these new possibilities, or is doomed to fade away.

Contents:

1. Introduction

Receptor binding is a technique which measures the *affinity* of a substance for a receptor. This affinity (from the Latin *affinitas*: 'attraction to', 'liking for') directly translates into biological activity. Receptor binding will never replace in vivo pharma-cological experiments, and it will never be possible to develop a drug solely on the basis of receptor binding data. But since the discovery of the opiate receptors by receptor binding, the scientific value of the technique has not been doubted. And within years it had firmly established its role in drug discovery. During the last two decades there has been an exponential increase in knowledge about different receptor (sub)types and in availability of radioligands. Many of the individual steps of the receptor binding technique have been at least partially automated. These facts, taken together with the indisputable advantage of the technique, the generation of highly accurate and specific data on potency at receptors, are the main reasons why receptor binding is currently employed at such a large scale, especially in industry.

The most important use of receptor binding is still straightforward determi-nation of the affinity of a compound for a specific receptor. Until detailed knowledge about receptor structure becomes available, large scale screening is the only possible way to detect primary lead compounds. Subsequently, quantitative structure affinity relationships form the basis of synthetic endeavors to arrive at optimal drugs.

The determination of a 'complete' receptor binding profile of a pharmacological interesting compound is one of the most exciting and rewarding possibilities in drug discovery. Apart from establishing its affinity for the receptor important for its pharmacological activity, the binding profile reveals much about the selectivity of the compound, and about its side-effect potential..

Receptor binding techniques have a broader potential than mere affinity determinations. The effects of long term drug treatment can be studied by 'ex-vivo' binding techniques, employing tissues of treated animals. Such studies have proven useful in the explanation at a molecular level of processes like tolerance and addiction, and in the prediction of the potential occurrence of such phenomena with new compounds. Receptor binding techniques are extremely powerful tools in the elucidation of the mechanism of action of drugs developed on the basis of functional tests, and in the determination of the potential side effects of drugs. The latter process, investigation of potent compounds in an as large as possible series of different binding assays has frequently led to the quasi-serendipitous discovery of new leads as well. Although the impact of receptor binding in the areas of drug metabolism and pharmacokinetics hitherto has been rather modest, there are several examples of the usefulness of the technique in these areas, too.

2. Some Technical Aspects of Receptor Binding, Definitions

Receptor binding essentially involves bringing together three components: a *receptor*, a *radioligand* (a pharmacological active compound labeled with either [^{3}H] of [^{125}II]), and a '*testcompound*' (a substance one wishes to investigate). The three components are brought togetheri n an *incubation mixture* consisting of a buffersystem keeping the mixture at a specific pH (mostly in the range 7.0 - 8.0), different ions, protease inhibitors, etc. Incubations are always carried out at a specific temperature (in the range between 0° and 37°C). Incubation times may vary from ten minutes up to several hours. For each different assay optimal incubation conditions have to be established.

During the incubation both the radioligand as well as the testcompounds come in the vicinity of the receptor, and are attracted by electrostatic forces. If affinity exists, the molecules will adept the most ideal configuration, and bind to the receptor via ionic forces, hydrogen bonds and/or Van der Waals forces. Usually this binding is reversible, but for instance in case of alkylating compounds, also irreversible covalent binding is possible. When, under the right conditions, equilibrium has been reached, the incubation is stopped by either filtration or centrifugation. Both processes separate free radioligand from bound, and the latter is consequently measured by liquid scintillation counting (LSC), from which in a straightforward way concentrations can be calculated. The radioactivity found on the filters from the incubations with the radioligand only (without any testcompound) is termed: '*total binding*'. Residual radioactivity found in incubations in which a large excess (100- or 1,000-fold) of a non-radioactive substance with affinity for the same receptors was present is termed: '*non-specific binding*'. The difference between the two is the '*specific binding*'. The displacement of a test-compound is determined at different concentrations. For each the binding is measured, and expressed as percentage of specific binding, thus:

$$B_{specific} = B_{total} - B_{non\text{-}specific} \quad \text{and} \quad \frac{B_{testcompound} - B_{non\text{-}specific}}{B_{specific}} \times 100\%$$

From these percentages a displacement curve can be constructed, and from this the IC_{50}, the concentration giving 50% displacement, can be extrapolated.

When specific binding is observed, it is certain that *specific binding sites* are present. This term however, is not synonymous with the term *receptor*. Every receptor is a specific binding site, but a specific binding site is a receptor only if it meets a number of criteria. First of all there has to be an *endogenous ligand*, and the interaction of this ligand with the specific binding site has to be *reversible*. Binding also has to be *saturable*: a given amount of tissue contains a given number of receptors only, and by incubating increasing concentrations of radioligand, at a certain concentration all receptors are occupied, and the specific binding should remain at a constant level. Next, binding should be *specific*, meaning that only compounds with the same pharmacological activity (or the reverse, since binding experiments do not discriminate between agonists and antagonists) should be able to produce displacement of the radioligand. Furthermore, the *kinetics* of the binding should match reality: binding must be temperature dependent, and at 37°C equilibrium should be reached at least as fast as the in vivo pharmacological effect becomes evident. In order to be termed receptors, specific binding sites should have a *relevant localization*: they must be present in tissues in which the endogenous ligand has a pharmacological effect. And finally, if agonists or antagonists at the receptor in question show stereoselectivity, this phenomenon should also be reflected in the affinities of the enantiomers of these compounds.

Displacement studies result in the IC50-value of a compound. Usually, this value is then transformed into the K_i-value, the compound's dissociation constant, via:

$$K_i = IC_{50} / (1 + {}^{S}/_{Kd}) \qquad \text{(the Cheng-Prusoff equation)}$$

In this equation S and Kd represent the concentration and the equilibrium dissociation constant of the radioligand respectively. IC_{50}- and K_i-values are expressed in molar concentrations, or often as their negative logarithms: pIC_{50} or pK_i, in order to allow direct comparisons with for instance pA_2-values from functional experiments. The mathematics of receptor binding are simple in many cases, and even the most complex calculations, those in which two or more binding sites are involved, are relatively simple for modern computers programmed for curve fitting. In many instances data obtained from binding experiments, that is LSC data, are directly fed into a computer which is programmed to execute the necessary calculations. After visual control of the affinity data generated, they are stored in databases.

2.1 SOURCES OF RECEPTORS, RADIOLIGANDS AND TESTCOMPOUNDS

Until the early nineties, *sources of receptors* were mostly tissues taken from laboratory animals, notably rodents, or from animals slaughtered for their meat. At present, these sources still account for a substantial part of the receptors used in binding assays. The other part consists of genetically engineered cells containing very specific receptors only, quite often from human origin. The preparation of receptor suspensions usually is rather straightforward, yet one of the aspects of receptor binding which has not been automated. The first step usually is homogenization. Dependent of the structure of the tissue involved, in this process a Polytron (blender) and/or Potter are used. After homogenization the further procedure involves a series of centrifugation steps, finally resulting in a preparation containing membrane fragments with intact receptors. These preparations can either be used immediately, or frozen (at -80°C) for future use.

Radioligands in most cases are bought from commercial suppliers, of which Amersham and New England Nuclear are the most important. Both companies, as well as a number of smaller ones, also offer possibilities for 'custom synthesis'. And of course one can also do the synthesis of radioligands in-house.

There are many possible *sources of testcompounds*. In fact any pure compound or mixture, synthetic or from a natural source may be used. If it contains anything with affinity for the receptor in question, it will be detected. Until recently, sample preparation was often a bottleneck in large scale screening operations. Each compound had to be taken from storage, a minute quantity had to be accurately weighed, to be dissolved, to be diluted, and to be pipetted into testubes or 96-well plates. This whole process can now be automated (at very high expenses!), but instead many industries have chosen for the strategy to prepare 96-well plates containing high concentrations (e.g. 10^{-2} M) of DMSO-solutions of testcompounds. These 'motherplates' can generate numerous 'daughterplates' with lower concentrations, and from these again large numbers of testplates can be generated. The process of transferring an entire compound collection into 96-well plates is very laborious, but after that one can make use of the solutions in hundreds of different assays!

2.2 SELECTIVITY, SENSITIVITY AND CAPACITY OF BINDING ASSAYS

Receptor binding assays are very *selective* in the sense that they concern a single molecular mechanism only. This selectivity is obtained by using radioligands which have a high affinity for a single population of receptors only. If such a ligand is not available, sometimes a 'selective tissue' is available, containing only one population of certain receptors. Modern biotechnology in many cases can provide cells in which receptors have been cloned which do not naturally occur in those cells. Finally, a last possibility is to block interfering receptors by selective unlabelled compounds.

Binding assays are *sensitive*. In case of very potent compounds concentrations in the picogram per milliliter range can be determined. Until recently this sensitivity was way beyond the range of any other analytical technique. The factor that determines this sensitivity is of course the radioactivity: [^{3}H]-ligands usually have specific activities in the range between 10 and 100 Ci/mmol, whilst [^{125}I]-labeled compounds have a specific activity of 2,200 Ci/mmol. Picomol or even femtomol quantities of such ligands can still be reliably measured. Theoretically, the affinities of unlabeled compounds range between zero and the inverse of Avogadro's number. In reality, compounds are mostly termed 'inactive' if their affinity is below 10^{-5} M (10,000 nM).

Potent' compounds have affinities ranging from 10^{-6} - 10^{-9} M (1,000 - 1 nM), whilst 'very' or 'extremely' potent compounds have affinities below 10^{-9} M (below 1 nM).

Binding assays have a *'high' capacity*, just like the term 'potent' not strictly defined. In the early years of receptor binding capacity was comparable to that of other in vitro techniques: several hundreds of samples (incubations) per day (for two analysts). Gradually, automation using sophisticated computer driven robots increased the capacity by an order of magnitude, still using single tubes. The introduction of 96-well plates (8 rows, 12 columns), combined with ever improving cell harvesters further boosted capacity by another order of magnitude. In recent years, receptor binding assays have been completely automated, and all manual actions have been taken over by a robot. Such laboratories in theory can run 24 hours per day, and in weekends, too! Also in recent years new techniques were developed: techniques no longer requiring separation of bound from free ligand by filtration. These techniques, termed 'homogeneous' or 'mix and measure' assays, and of which Amersham's 'Scintillation Proximity Assay' (SPA) and New England Nuclear's

'Flashplate' are the first examples, again raised capacity. Miniaturization is yet another trend: already the 384-well plate (16 by 24 format, essentially four 96-well plates on the surface of one) has been used for binding assays with the SPA technique. These plates can be measured (radioactivity, fluorescence) by Packard's Topcount®. Even plates with higher numbers of wells have already been made: 864 (24 x 36, nine 96-well plates on the surface of one). Currently pipetting robots and analytical instruments are being developed which can handle these plates. Next to miniaturization, the latest trend is the development of non-radioactive assays. Time Resolved Fluorescence (using lanthanide labels such as Europium) are the first examples of those.

At this moment, under optimal conditions, fully robotic high throughput screening laboratories can produce approximately 50,000 samples per day. And it may be expected that future developments will further increase that number.

3. Different Forms of Receptor Binding: in vitro, ex-vivo and in vivo

By far most of the receptor binding work concerns straightforward *in vitro* affinity determinations, meaning that the source of receptors is taken from laboratory animals, from slaughterhouse material, human biopsy or autopsy tissue, or cells grown in culture. Another type of binding experiment is the determination of a 'saturation curve' This implies incubation of increasing concentrations of labeled compound with a fixed tissue concentration (a fixed number of receptors). From these experiments one obtains both the number of receptors present in a certain amount of tissue ('B_{max}') as well as the equilibrium dissociation constant ('K_d'). Because saturation curves are very often analyzed by a method first described by Scatchard, they are often referred to as 'Scatchard plots'. A third type of in vitro binding are 'association' and 'dissociation' experiments: curves obtained by incubating a fixed amount of labeled compound with a fixed number of receptors for varying time intervals. Data generated this way are association and dissociation rate constants respectively. In the first place an alternative route to the equilibrium dissociation constant K_d (which is the quotient of both constants), and secondly of importance in pharmacokinetic studies.

Ex-vivo receptor binding is a hybrid between in vitro and in vivo binding. It concerns binding experiments using tissue of treated animals compared with untreated controls. Purpose of this type of studies usually is to investigate the effects of acute or chronic drug treatment, or the effects of either surgical or chemical lesions on certain receptor populations. In most cases these studies concern saturation experiments. Observed differences in the K_d-value of the radioligand are interpreted as alterations in receptor structure; changes in receptor numbers are indicative of either receptor upregulation, downregulation or effects on receptor dynamics: shifts in diurnal (circadian) or seasonal receptor rhythms. Generally speaking chronic administration of receptor agonists results in receptor downregulation ('desensitisation') and chronic administration of receptor antagonists in receptor upregulation ('sensitisation'). Both processes may lead to unwanted side effects of chronic drug treatment, and both may result in more or less severe withdrawal symptoms after cessation of drug treatment. In certain cases it has been shown that receptor binding experiments on chronically treated animals have a distinct predictive value for the occurrence of such phenomena in patients.

In vivo receptor binding is a term comprehending experiments in which radio-labeled compounds with a high specific activity are administered to living animals. Compared to in vitro experiments, in vivo binding is expensive (it requires large amounts of labeled compounds), laborious (pipetting into a test-tube can be automated, drug administration to animals not), and it produces large amounts of radioactive waste. This notwithstanding, in vivo receptor binding techniques will generate data which can not be obtained in vitro. It is an unambiguous way to study receptor localization and receptor densities,and also produces pharmacokinetic data: penetration of special tissues like brain fetus, etc. In vivo binding is a powerful tool in unraveling vivo/vitro discrepancies as well, and when the compound under investigation is available in radiolabeled form, in vivo binding experiments may simultaneously generate data on metabolic processes.

3.1 ELUCIDATION OF THE MECHANISM OF ACTION OF A DRUG

When a compound is developed on the basis of a certain mechanism of action, actual proof of that mechanism is usually uncomplicated. The compound will show a clear affinity for the receptor in question, and there will be a straightforward correlation between that affinity and a certain biological activity. Final proof of the mechanism of action can be obtained by labeling the drug and proving that it binds to the receptor in question by both conventional binding assays as well as by autoradiography.

When the mechanism of action of a compound is unknown, receptor binding is a very likely technique to assist in its elucidation. Firstly, the compound should be screened in all available assays. In such cases the possibilities of contract houses like Novascreen or Panlabs should not be overlooked. If the compound shows affinity in one or more assays, these should be verified by testing of other compounds with affinity for the same receptors. If a correlation is found between a certain affinity and a functional pharmacological test, the problem is very likely solved. If the compound does not show affinity for any known receptor, one has to proceed by labeling the compound and to investigate its binding properties by conventional saturation and displacement studies as well as by autoradiography. These studies may result in the identification of the receptor in question, which may be one hitherto undescribed.

3.2 DRUG METABOLISM, PHARMACOKINETICS AND RECEPTOR BINDING

Because drug metabolism studies are laborious and expensive, they are usually only performed on compounds in late stages of development. Information on metabolism however, is valuable in an early stage of drug development as well. Also here, binding studies may be of help. If one incubates a compound (with a known affinity) with a liver homogenate and samples after different times, one can measure the affinity of the samples in the appropriate assay. If it either increases or decreases, an indication of metabolic activation or deactivation is obtained. More likely than not, it will prove to be simple to extract the metabolites from the incubation mixture, to determine their structure, to synthesize and to test them. Metabolites detected this way are very likely to be metabolites in vivo as well.

The extrapolation from an in vitro affinity for a certain receptor to efficacy in patients is full of uncertainties. One of them is pharmacokinetics. With receptor binding techniques,in some cases valuable information on different aspects of pharma-cokinetics can be obtained. The only condition is that the mechanism of action of the drug in question is known. With in vivo binding penetration of the blood brain barrier can be established unambiguously, and also data on duration of action can be obtained.

In more traditional aspects of pharmacokinetics: determination of plasma levels, receptor binding may be useful, too. If the drug in question is a highly potent compound with a known mechanism of action, a binding assay can be used to quantify 'plasma affinity' rather than the determination of the parent compound only (as with conventional analytical methods), With binding the affinity of the parent compound with its metabolites is determined. Especially in cases of compounds which generate active metabolites, measurement of plasma affinity correlates better with duration of pharmacological effects than the plasma concentration of the parent compound.

3.3 LIMITATIONS: 'FALSE' POSITIVES AND 'FALSE' NEGATIVES

Contrary to the popular belief, there is no such thing as 'nonsense binding', binding which does not mean anything. Even the generally accepted term 'non-specific binding' is nothing more than an operational euphemism covering ignorance. Receptor binding by itself will never result in 'false' results, neither negative nor positive. It is only when one compares different techniques that these terms have pragmatic value.

'False positives': compounds active in binding experiments but inactive in vivo, are merely a nuisance, no threat in drug discovery. They are usually ignored, but if one investigates such cases, explanations are invariably found in either of the following possibilities: 1) experimental errors; 2) reversed activity: if one screens for antagonists starting with binding, agonists will be 'false positives' and vice versa; 3) the compound is not absorbed from the gut, is rapidly metabolized, or does not cross the blood-brain barrier; 4) the compound has other activities, more potent than the one studied, or counteracting it.Cardiac glycosides for instance, have a low affinity for benzodiazepine receptors, but this can not be demonstrated in vivo because the animals die of cardiac arrest long before these drugs reach the necessary concentrations to have an effect on benzodiazepine receptors.

'False negatives': compounds negative in binding experiments, but active in vivo, are a definite risk in drug discovery. Whenever one starts screening with a binding assay one should realize this. False negatives invariably fall in either of the following categories: 1) experimental errors; 2) 'prodrugs': compounds metabolized to active compounds; 3) compounds acting by a different mechanism. This is the only serious drawback of screen-

ing by specific receptor mechanisms. Many pharmacological effects, like e.g. analgesia, hypotension or inhibition of acid secretion can be evoked by a number of different molecular mechanisms. When using an in vivo functional test, all known and unknown receptors are 'open' to the compounds tested. When using a single binding assay, one deliberately restricts the finding of active compounds to a single mechanism of action.

4. Why Receptor Binding with Extracts from Natural Sources?

Nature not only is the source of food, it also has provided mankind with numerous industrial products, oils, flavors, vitamins, spices, cosmetics, pesticides and DRUGS!

Most certainly not only ill-defined herbal remedies and homeopathic preparations, but also prescription drugs. Flowering plants and ferns have produced approximately 120 commercially sold drugs. In 1992, in the USA, 13 out of the top-25 best selling drugs were from natural sources, and accounted for a full 25% of all drug prescriptions! These drugs include a wide variety of chemical classes, e.g.: aliphatics (phosphonomycin), alicyclics (lovastatin), amino acid derivatives (penicillins, cyclosporins), aromatics (griseofulvin), carbohydrates (deoxynojirimycin), macrocyclic lactones (avermectins), N-containing heterocyclics (asperlicin), O-containing heterocyclics (podophyllotoxin), quinones (tetracyclins), etc.

It would be a mistake to assume that natural sources have been more or less exhaustively studied. The contrary is true: of the estimated number of 265,000 plant species for example, an ample 70% have not been investigated at all. The other 30% has only been partially explored, and less than 0,5% has been exhaustively studied, at least their chemical composition. There hardly has been a systematic research of any natural source in terms of molecular mechanisms of action.

5. Results of Binding Experiments with Extracts from Natural Sources

Except for cases in which one is specifically searching for a certain class of compounds, generally speaking one has no idea about the chemical nature of the biologically active compounds one is screening for. And thus, there is no way of knowing whether or not a given extraction procedure is the right one. Extensive binding studies with different species extracted by different solvents such as water, ethanol, chloroform, toluene and dichloromethane/methanol mixtures learned that the results may vary considerably, not only qualitatively, but also quantitatively. This therefore, is a strategy which is generally applicable.

Although plant cell cultures may not be expected to synthesize compounds which are not present in the plant under physiological conditions, major differences between extracts from e.g. leaves and cell cultures have been observed, both quantitative as well as qualitative, albeit that the latter may simply be a matter of differences in concentration.

Closely related plants may differ considerably in their spectrum of secondary metabolites, as is reflected in their binding profiles, and in reverse: unrelated plant species may contain compounds acting on the same receptors.

As far as pure natural compounds are concerned: they of course are in no way different from synthetic compounds in general. Natural compounds are not by definition more potent than synthetic compounds, and neither more selective. There are numerous examples of extremely potent and selective natural compounds (e.g. atropine, morphine, d-tubocurarine), but there are also many examples of potent natural compounds which have comparable affinities for a whole range of receptors. The ergot alkaloids belong to this class.

6. From a Plant to a Drug: Main Questions, Strategies

In drug research there is only one problem: finding the mechanism by which a drug has to act in order to achieve a therapeutic effect in a given disease, or ultimately, finding the mechanism by which a drug has to act in order to eliminate the cause of the disease. Once such mechanisms are identified, the rest will be history in a relatively short time. Because there are no apparent strategies which will lead to the unambiguous identification of the mechanisms mentioned above, drug discovery will continue to be a development of potential drugs acting on specific receptor systems which will have to be evaluated in the clinic.

Once a theory on a certain mechanism of action has been decided upon, the first step is lead finding. In that process natural sources will play a role next to conventional and combinatorial chemistry. Once an extract has been identified which shows activity in the binding assay in question, the next step is to isolate the active compound. For this, the same assay should be used. After purification and structure elucidation there are two main scenarios. First, the compound in question is relatively simple, and easy to synthesize. In that case, the logical next steps are the synthesis of derivatives, and the selection of the best compound for further development. In reality however, with extracts from natural sources a second scenario is more likely: the active compound will turn out to be 'impossible' to synthesize on a commercial scale. Then several options remain: the compound can be extracted from the original source, either naturally occurring or from cultivations, or one could try to establish a cell culture as a source. From this it follows that it is a good strategy to start screening operations with extracts from natural sources which are either abundant, already cultivated, or of which stable cell cultures already exist.

A second kind of question which occurs with natural sources, notably those of which an (ethno)pharmacological activity or toxicity is known, is: "what is the possible mechanism of action of this plant, extract or compound?" To answer such questions one has to use as many as possible different binding assays. In such cases the best option is to make use of contract houses which offer a great variety of different binding assays. Examples of such companies are Novascreen and Panlabs in the USA, Neurotech and Batelle in Switzerland, and Cerep in France. The facilities offered by the contract houses may be costly, but the alternative: to set up several hundreds of different assays is infinitely more expensive, if at all feasible.

7. Bindingstudies on Plant Receptors: a Challenge!

So far, little if any receptor binding studies have been carried out on plant receptors. There are no *a priori* arguments against the use of plants as sources for receptors in binding studies. Question is: why would one do that? If one is willing to accept the hypothesis that everything in nature has a purpose, it follows that all plant constituents, including 'secondary metabolites' must have a function, either inside the plant or in some of the species in the same ecosystem. Thus, it is very likely that plants contain receptors for some of the chemicals they synthesize. This hypothesis is very easy to verify: obvious possibilities are for instance to investigate Papaver species for the presence of opiate receptors, Belladonna species for muscarinic receptors and Claviceps species for monoamine receptors.

If the concept proves to be true, it would open alternative strategies in the search for biologically active compounds. If the search for ligands for the "X" receptor for instance, would remain unsuccessful, one could try to search for plants which contain that receptor. If such a plant is found, it is bound to contain endogenous ligands as well! Alternatively, if this route fails, one could for instance search insect species for the presence of the receptor in question. If such an insect if found, it is very likely that plant species on which they feed, *or which they actively avoid*, contain natural ligands for that receptor.

8. Ethnopharmacology: Barking up the Wrong Tree?

Ethnopharmacology continues to be an important starting point in the area of drug research on natural products. It is questionable however, whether this obvious inroad deserves the attention it is given. Knowledge about the (dis)functioning of living organisms has largely been obtained in the last century. Before that, mankind was relatively ignorant, not aware of the mere existence of for instance bacteria, viruses and receptors. Traditional medicines have been found by coincidence, or at best by some trial and error, but most certainly not by systematic research. It may therefore safely be assumed that 99% of all valuable drugs present in nature have been missed.

One example to illustrate the statement that it is extremely unlikely that primitive civilizations did anything in the direction of systematic research are the arrow poisons. The notorious 'curare' has been used by many tribes in South America. It is noteworthy that different types of the poison were extracted from plants with quite different appearances, belonging to different families. Also the active substances belong to different chemical classes, yet they all have in common their mechanism of action: neuromuscular blockade. Despite the fact that the South American forests are a rich source of plant species which contain cardiac glycosides, none of these has ever been used as arrow poison. The reverse applies to the African continent where many tribes used plants containing cardiac glycosides as sources of extremely powerful arrow poisons, but none have discovered any of the abundant plant species containing curare-like substances. Plants containing aconitine or related alkaloids, also have been used as arrow poisons, but exclusively in some Asiatic countries. Remarkably, substances like batrachotoxinine, chemically quite different but acting on the same ion-channel as aconitine, are found in the skins of a number of species of tree-frogs used as sources of poisons for blow-pipe arrows in Central America. Many more examples can be given, but the point should be clear. Based on what we know now about toxic substances in plants, none of us would have to leave his own backyard in order to prepare a perfectly lethal arrow poison. Therefore, the conclusion is that ethnopharmacology is *one* road, but certainly *not the only one*, and most likely *not the best one*.

9. Chemical and Biological Diversity: no Holy Grail!

The call for 'chemical diversity' in screening operations seems to be universal. From a purely theoretical basis this seems to be logical, but the concept that chances for success in leadfinding are improved by the strive for chemical diversity is not generally applicable. Arguments for this statement are the bewildering chemical diversity of *natural* compounds acting on for instance nicotinic receptors: from very simple (nicotine, cytisine, anabaseine, etc.) to very complex (erythroidine, methyllyaconitine, d-tubocurarine, peptides, etc.). And the fact that many useful drugs, either natural (e.g. ergotamine) or synthetic (e.g. clozapine) have affinities for many different receptors. In theory, those compounds could serve as leads for all these different receptors! Such examples of 'intramolecular chemical diversity' should put the quest for chemical diversity in proper perspective!

Likewise, in natural product research, there is a persistent call for biological diversity. Again, this seems to be logical, but the concept that chances for success in leadfinding are improved by the strive for biological diversity is not generally applicable either. The argument to support this statement is that many completely unrelated living organisms contain ligands for e.g. nicotine, muscarine, or opiate receptors. If a mechanism (neurotransmitter) is of any importance, as a result of that it is bound to be present in many different living species.

In summary: there is nothing wrong with the strive for either chemical or biological diversity, but it is definitely no holy grail! In the absence of knowledge about receptor structure or ligands other than the endogenous, leadfinding is a numbers game!

10. Combinatorial Chemistry: a Threat to Phytopharmacology?

Until the early nineties, new compounds were exclusively synthesized one by one, purified, crystallized, analyzed and submitted for testing. Often they were prepared in fairly large quantities: grams, or at least several hundreds of milligrams. For large scale screening operations companies relied on their stocks of tens-, or even hundreds of thousands individual compounds.

But since then, a new philosophy is developing at a staggering speed: *'combinatorial chemistry'*: the synthesis of extremely large numbers of different compounds, either in mixtures or as single compounds. Quantities are usually very small: milligrams or less; purification is crude: quite often one settles for something like 80%; and analysis is performed on only a fraction of the compounds. The syntheses of "libraries" of small peptides were the earliest examples of combinatorial chemistry, but soon attention was directed towards more stable 'peptoïds' and other compounds.

In the last few years, in combinatorial chemistry the balance is shifting from synthesizing libraries towards single compounds in a process named 'high speed parallel synthesis'. Starting from a more or less complicated core-molecule, many different substituents are attached step by step, thus resulting in large numbers of different single compounds which are directly transferred into 96-well plates.

As a result of the different new approaches collected under the term "combinatorial chemistry" the output of synthetic departments has been increased by several orders of magnitude. These new possibilities, not only in the lead finding process, but also in lead optimization, form a serious threat to all other sources of compounds which are less easy accessible. This is especially true for complicated mixtures such as extracts from natural sources, from which biologically active compounds have to be identified, isolated and purified before structure elucidation can even begin. After that, there still remains the need to isolate the compound in milligram quantities, or to synthesize it.

Despite the fact that virtually everybody has the gut-feeling that Nature contains all the important lead compounds we will ever need, the combination of the possibilities of combinatorial chemistry and high throughput screening are so alluring that in all likelihood it will take at least a decade before it will be realized this may not be the answer to everything.

BIOLOGICAL AND CHEMICAL EVALUATION OF PLANT EXTRACTS AND SUBSEQUENT ISOLATION STRATEGY

A. MARSTON and K. HOSTETTMANN*
Institut de Pharmacognosie et Phytochimie
Université de Lausanne
B.E.P., CH-1015 Lausanne, Switzerland

More than 85% of higher plants have not been adequately surveyed for potentially useful biological activity and the plant kingdom has not received sufficient attention as a resource of possible medicinal agents. At the same time, it has been estimated that over half of the world's 25 best selling pharmaceuticals for 1991 owed their origin to a natural source material (O'Neill and Lewis, 1993). Another statistic is that approximately 120 plant-derived chemical compounds are currently used as drugs. Many of these are extracted and purified directly from plants (Farnsworth *et al.*, 1985). A recent success story is the market for *Ginkgo biloba* (Ginkgoaceae) extracts in disorders of peripheral blood circulation and geriatry. Annual sales in Europe are estimated at US $ 500 million (Sticher 1993).

The conclusion to be drawn is that much more work needs to be done on the investigation of plants (and other sources of natural products) in the search for novel bioactive compounds. These may be useful as they are or they may form the basis of new lead compounds for further exploitation.

In view of the large number of plant species potentially available for study, it is essential to have efficient systems available for their rapid chemical and biological screening.

1. Chemical Screening

Isolation of pure, pharmacologically active constituents from plants remains a long and tedious process. For this reason, it is necessary to have methods available which eliminate unneccesary separation procedures. Chemical screening is thus performed to allow localization and targeted isolation of new or useful types of constituents with potential activities. This procedure enables recognition of known metabolites in extracts or at the earliest stages of separation and is thus economically very important.

Thin-layer chromatography (TLC) is the simplest and cheapest method of detecting plant constituents because the method is easy to run, reproducible and requires little equipment (Marston *et al.*, 1997). However, for efficient separation of metabolites, good selectivity and sensitivity of detection, together with the capability of providing on-line structural information, hyphenated high performance liquid chromatographic (HPLC) techniques are preferred (Hostettmann *et al.*, 1997a). They play an important role as an analytical support in the work of phytochemists for the efficient localization and rapid characterization of natural products.

HPLC coupled to a UV photodiode array detector (LC/UV) has been widely used for the analysis of crude plant extracts. The UV spectra of natural products obtained on-line by LC/UV give useful information on the type of constituents and in the case of certain classes of compound, such as the polyphenols, indications of oxidation patterns (Hostettmann *et al.*, 1994).

HPLC coupled to mass spectrometry (LC/MS) has been introduced recently and is still little used by phytochemists (Wolfender *et al.*, 1995). Mass spectrometry is one of the most sensitive methods of molecular analysis and yields information on the molecular weight as well as on the structure of the analytes. Due to the high power of mass separation, very good selectivities can be obtained. However, it has been difficult to achieve on-line coupling of HPLC and MS. These problems have now been overcome with the introduction of different LC/MS interfaces. For the HPLC screening of crude plant extracts, three interfaces have been used in our laboratory: thermospray (TSP), continuous flow fast atom bombardment (CF-FAB) and electrospray (ES). They cover the ionization of relatively small non-polar products (aglycones, MW ca. 200) to highly polar molecules (glycosides, MW ca. 2000). LC/TSP-MS allows satisfactory ionization of moderately polar constituents such as polyphenols or terpenoids in the mass range 200-800 amu. For larger, polar molecules such as saponins (MW > 800), CF-FAB or ES are the methods of choice (Wofender *et al.*, 1995).

HPLC coupled to a NMR spectrometer (LC/NMR) has, until recently, been little used mainly because of its lack of sensitivity. However, recent progress in pulse field gradients and solvent suppression, together with improvements in probe technology and the construction of high field magnets now allow many applications of the technique. LC/NMR has important potential for on-line structure identification of natural products. Indeed, NMR spectroscopy is by far the most powerful spectroscopic technique for obtaining detailed structural information about organic compounds in solution (Albert 1995). Coupling to a HPLC instrument is straightforward and solvent suppression techniques now allow the use of non-deuterated solvents (methanol or acetonitrile) under reversed-phase conditions. Water is replaced by D_2O.

2. Biological Screening

Screening programmes for biologically active natural products require the right bioassays. Detection of compounds with the desired activity in complex plant extracts depends on the reliability and sensitivity of the test systems used. Bioassays are also essential for monitoring the required effects throughout activity-guided fractionation: all fractions are tested and those continuing to exhibit activity are carried through further isolation and purification until the active monosubstances are obtained. The search for promising plant extracts and subsequent activity-guided isolation put specific requirements on the bioassays to be used. They must be simple, inexpensive and rapid in order to cope with the large number of samples - including extracts from the screening phase and all fractions obtained during the isolation procedure. They must also be sensitive enough to detect active principles which are generally present only in small concentrations in crude extracts. Their selectivity should be such that the number of false positives is reasonably small (Hostettmann *et al.*, 1995).

In this chapter, emphasis will be placed on TLC autographic assays, which combine TLC with a bioassay in situ and allow localization of active constituents in a complex matrix (Hostettmann *et al.*, 1997b).

The number of available targets for biological screening is limited. Furthermore, bioassays are often not reliaby predictive for clinical efficiency. For these reasons, it is extremely helpful to have chemical screening techniques available as a complementary approach for the discovery of new molecules which might serve as lead compounds. As mentioned above, chemical screening also serves for dereplication purposes.

2.1. TLC SCREENING FOR ANTIFUNGAL AND ANTIBACTERIAL COMPOUNDS

The use of immunosuppressive drugs and the spread of AIDS have resulted in an increasing occurrence of opportunistic systemic mycoses. The infections commonly observed in the immune-compromised host include candidiasis (*Candida albicans* and related species) of the oesophagus and mouth, cryptococcosis (*Candida neoformans*) and aspergillosis (*Aspergillus flavus*, *A. fumigatus*, *A. niger*). As there are few really effective antifungal preparations currently available for the treatment of systemic mycoses and as the efficacy of existing drugs is rather limited, it is important to find new sources of antifungal agents. Plant-derived natural products may offer potential leads for novel agents which act against these mycoses.

There is also a need to screen plants for constituents which have activity against plant pathogenic fungi: fungal attack can be economically devastating in agriculture.

Bioautography is a very convenient and simple way of testing plant extracts and pure substances for their effects on both human pathogenic and plant pathogenic microorganisms. It can be employed in the target-directed isolation of active constituents. Three bioautographic methods have been described (Rios *et al.*, 1988):

- agar diffusion (in which the antimicrobial agent is transferred from the chromatogram to an inoculated agar plate through a diffusion process)
- direct bioautographic detection of the TLC plate (Homans and Fuchs, 1970; Hamburger and Cordell, 1987)
- agar-overlay (Rahalison *et al.*, 1991)

Direct bioautography is applicable to microorganisms that can grow directly on the TLC plate. The agar-overlay technique is a hybrid of the two other methods and is applicable to a broad spectrum of microorganisms. It produces well defined zones of inhibition and is not sensitive to contamination. Active compounds are transferred from the stationary phase to the agar layer (which contains the microorganism) by a diffusion process. After incubation, the plate is sprayed with a tetrazolium salt (e.g. MTT) which is converted to a formazan dye by the microorganism. Inhibition zones are observed as clear spots against a purple background.

Direct bioautographic procedures have been described for spore producing fungi such as *Aspergillus*, *Penicillium* and *Cladosporium* (Homans and Fuchs, 1970) and also for bacteria (Hamburger and Cordell, 1987). Numerous antifungal compounds have been characterized using *Cladosporium cucumerinum* in a routine assay (Hostettmann and Marston, 1990).

The agar-overlay assay has been used for yeasts such as *Candida albicans* and can also be applied to bacteria such as *Bacillus subtilis* (Rahalison *et al.*, 1991). If phenol red is incorporated into media containing 0.6% agar and the plates are sprayed with MTT, clearer results are obtained, with dark red coloured inhibition zones appearing against a blue background (Saxena *et al.*, 1995). This method works successfully with a range of microorganisms, including *Candida albicans*, *Escherichia coli*, *Pseudomonas aeruginosa* and *Staphylococcus aureus* (Saxena *et al.*, 1995).

2.2. TLC SCREENING FOR RADICAL SCAVENGERS AND ANTIOXIDANTS

Another use of TLC for biological testing is as a means for discovering new antioxidants in higher plants (Cuendet *et al.*, 1997). These can be detected on a TLC plate by spraying with 2,2-diphenyl-1-picrylhydrazyl (DPPH) radical. Antioxidants reduce the radical, producing white spots on a purple background. Alternatively, the bleaching of crocin (which normally gives a yellow colour on the plate) can be used to distinguish components of plant extracts with potential antioxidant or radical-scavenging properties.

3. Screening for Bioactive Compounds and their Subsequent Isolation

3.1. AN ANTIFUNGAL NAPHTHOQUINONE FROM *SWERTIA CALYCINA* (GENTIANACEAE)

Among the examples of natural products isolated in our laboratory using the TLC bioautographic approach is a 2-methoxynaphthoquinone from *Swertia calycina* (Gentianaceae), a small plant found in Rwanda. This example illustrates well the combined use of TLC and HPLC in the search for new antifungal metabolites (Figure 1). TLC bioautography of the dichloromethane extract of *S. calycina* showed the presence of a compound which strongly inhibited the growth of *C. cucumerinum*. HPLC-UV and HPLC-MS analyses of the extract revealed the presence of three main compounds: a bitter princi-

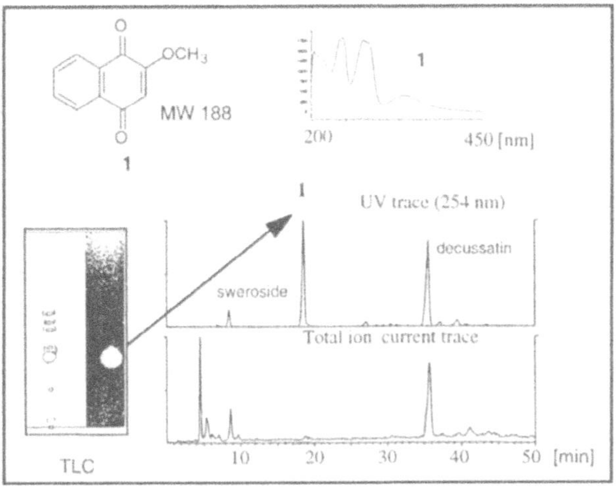

Figure 1. TLC bioautography *(C. cucumerinum)* and LC/UV/MS analysis of *Swertia calycina* (Gentianaceae) whole plant dichloromethane extract.

ple, a xanthone and a naphthoquinone derivative with a MW of 188. Comparison of on-line UV and MS data with a data bank allowed identification of the bitter principle as sweroside and the xanthone as decussatin. As these have no antifungal properties, the strong activity of the dichloromethane extract was attributed to the naphthoquinone, a class of compounds which is known to have strong antimicrobial properties. This was confirmed by TLC bioautography of the peak eluted from analytical HPLC (Figure 2). Targetted isolation

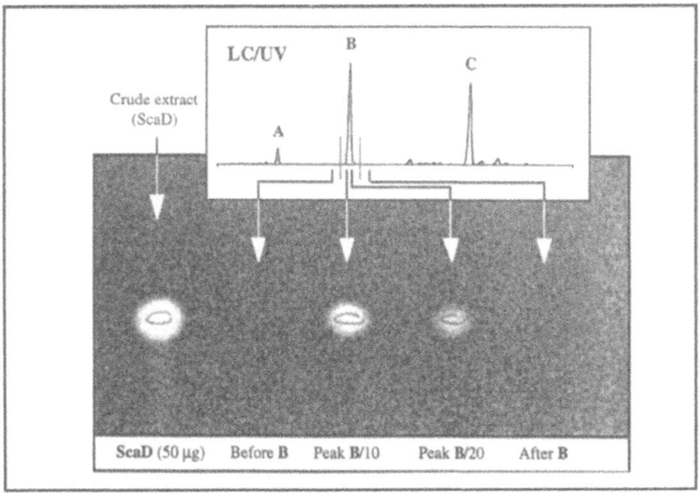

Figure 2. TLC bioautography *(C. cucumerinum)* of fractions collected from analytical HPLC of *Swertia calycina* whole plant dichloromethane extract (ScaD).

afforded the active compound, identified as 2-methoxy-1,4-naphthoquinone (1). Interestingly, quinones were previously not known to occur in the Gentianaceae. The minimum quantities of 1 required to inhibit the growth of *C. cucumerinum* and *C. albicans* on TLC plates were 0.1 and 0.4 mg, respectively (Rodriguez *et al.*, 1995). By comparison, the reference substance propiconazole was active at 0.1 and 0.001 mg, respectively.

3.2. ANTIFUNGAL CONSTITUENTS OF *OCOTEA USAMBARENSIS* (Lauraceae)

Ocotea usambarensis Engl. (Lauraceae) is a tree from central Africa which is commercially important for its resistant wood. The bark is used in traditional medicine for the relief of stomache ache and as a bitter in gastric complaints (Watt and Breyer-Brandwijk, 1962).

On preliminary bioautographic screening, the essential oil obtained by water distillation of the fresh bark of *O. usambarensis* from Rwanda exhibited strong antifungal activity against *Cladosporium cucumerinum*. Although gas chromatography coupled with mass spectrometry (GC-MS) showed a complex mixture of constituents, identification of the antifungal compounds was possible after preparative separation by overpressure liquid chromatography (OPLC).

Automated multiple development (AMD) chromatography with a multistep solvent gradient (t-butyl methyl ether-ethyl acetate-dichloromethane 5:3:20 in a 16-step gradient with 60-100% n-hexane) allowed localisation of the active constituents and was used to choose the isocratic eluent for OPLC separation. This latter was performed on a 200 x 200 x 2 mm Si 60 TLC plate with a Chrompres 25 OPLC instrument (Laboratory Instruments, Budapest, Hungary). The solvent chosen was t-butyl methyl ether-ethyl acetate-dichloromethane-n-hexane 5:3:20:72 and the flow rate was 1.8 ml/min. The sample size was 130 mg and the bands obtained after migration were extracted from the support with ethyl acetate-n-hexane 1:1.

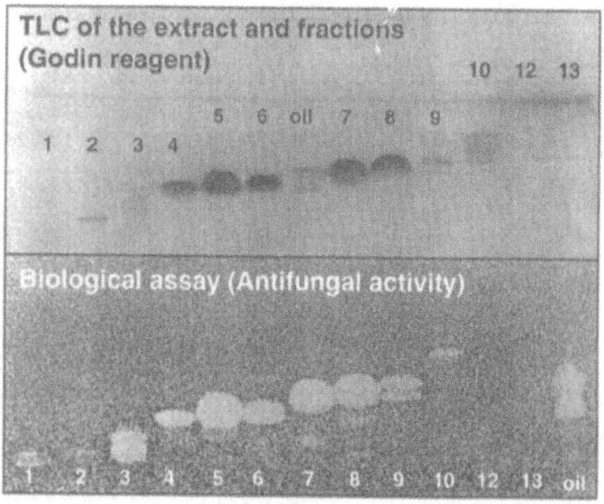

Figure 3. Analysis of fractions from *Ocotea usambarensis* (Lauraceae) after OPLC separation of the essential oil from the bark.

A total of 12 fractions was obtained and after analysis by AMD, the plate was sprayed with spores of *Cladosporium cucumerinum*. Incubation of the plate in a moist tank for 3 days gave the result shown in Figure 3 (lower chromatogram). Fungal growth was observed over the whole plate as a grey coloration, except in the lanes containing antifungal compounds. Fractions 3 to 9 had significant activity (Terreaux *et al.*, 1994).

Analysis of these fractions by GC-MS and bidimensional TLC revealed that the number of components of each fraction was much larger than could be observed on a normal TLC plate. An example of the analysis of fraction 4 is shown in Figure 4.

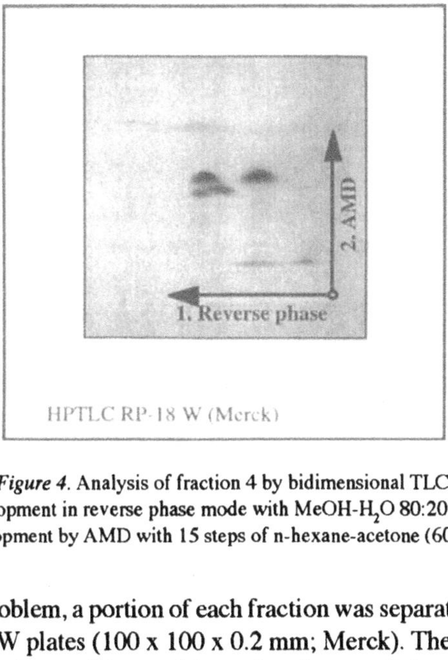

Figure 4. Analysis of fraction 4 by bidimensional TLC.
First development in reverse phase mode with MeOH-H_2O 80:20 (isocratic);
second development by AMD with 15 steps of n-hexane-acetone (60:40 to 90:10).

In order to solve this problem, a portion of each fraction was separated by two-dimensional TLC on HPTLC RP18W plates (100 x 100 x 0.2 mm; Merck). The first development was performed under isocratic conditions with methanol-water 8:2 and the second development with a 15-step AMD gradient of n-hexane and acetone (from 60% to 90% n-hexane). By running a reference bioautography plate, the active constituents could be localized, scraped from the TLC plate, extracted with n-hexane-ethyl acetate 1:1 and submitted to GC-MS analysis for identification purposes. This procedure for fraction 8 is shown in Figure 5.

From the GC-MS, the Kovats' indices and comparison with a reference standard, the active component of fraction 8 was identified as nerolidol (**2**). The minimum amount of this compound required to inhibit growth of *C. cucumerinum* in the TLC bioassay was 15 μg. The same procedure was adopted for the other fractions, in order to characterise the remaining antifungal constituents (Terreaux *et al.*, 1994).

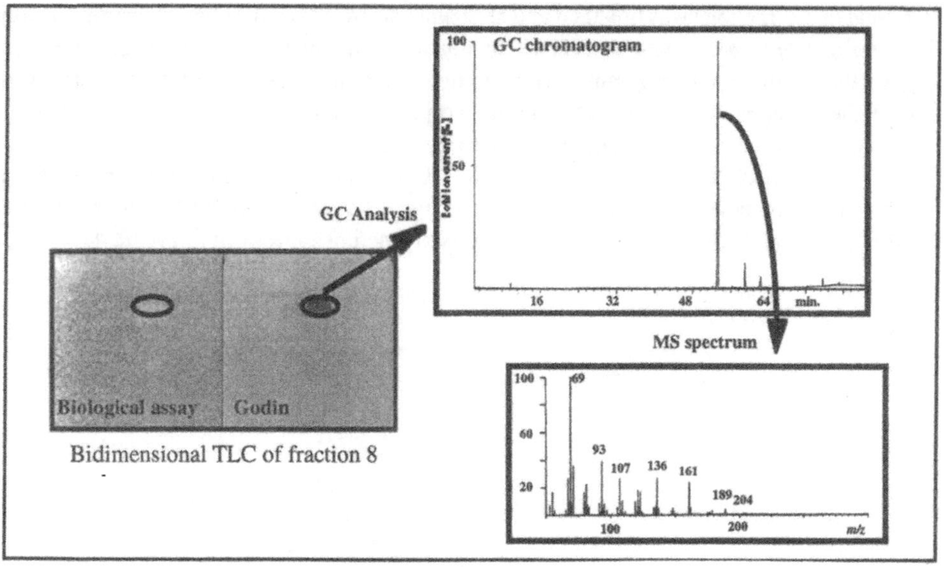

Figure 5. Identification of active constituents in OPLC fractions from *O. usambarensis* by 2D-TLC and GC-MS.

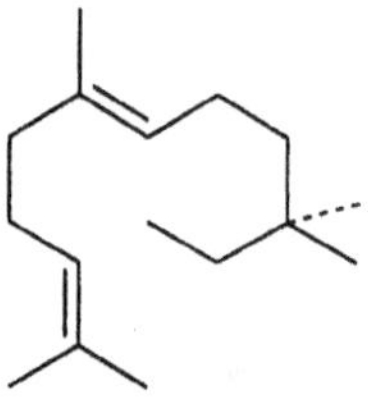

2

3.3. XANTHONES FROM *HYPERICUM ROEPERANUM* (GUTTIFERAE)

Considerable interest has recently been shown in the Guttiferae family, with special emphasis on the biological activities of the genus *Hypericum*. In particular, extracts of St. John's wort, *H. perforatum*, are now widely used in Europe for the treatment of depression.

A study was performed on the xanthones of *H. roeperanum*, a shrub growing in the southern parts of Africa. The dichloromethane extract of the roots gave a very complicated chromatogram after LC/UV analysis. For this reason, preliminary fractionation was carried out. Subsequent LC/UV/MS investigation using a thermospray interface showed the presence of xanthones in one of the fractions (Figure 6). As several of the xanthone peaks appeared to be novel compounds with molecular weights which did not correspond to previously described members of this substance class, further purification was undertaken.

The fraction proved very difficult to separate by the usual chromatographic techniques, including semi-preparative HPLC. However, the all-liquid technique of centrifugal partition chromatography (CPC) (Marston and Hostettmann, 1994) separated the xanthones into 6 fractions (Figure 7). Although these were not pure, the individual xanthones were well distributed among the fractions. All that was subsequently required was a final purification step (gel filtration, CPC or semi-preparative HPLC) to obtain the xanthones (**3-10**) (Rath *et al.*, 1996). Two of these, **9** and **10**, were new compounds.

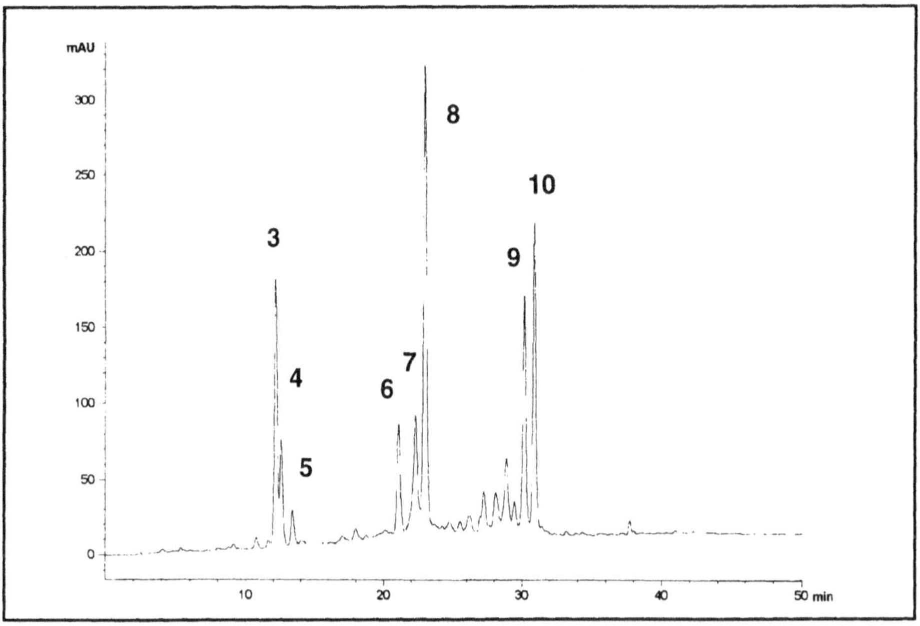

Figure 6. Analytical HPLC of the xanthone-containing fraction from a *Hypericum roeperanum* (Guttiferae) dichloromethane root extract.
HPLC: column Macherey-Nagel RP-18 (5μm; 4 x 250 mm); MeOH-H_2O 55:45 Æ 100:0 (with 0.05% TFA) in 40 min, then 100:0 for 10 min; flow-rate 1 ml/min; detection 254 nm.

The minimum amount of xanthones **6** and **8-10** required to inhibit the growth of *Candida albicans* in the TLC bioassay was 1 μg. All the other xanthones were inactive at 10 μg. This compared with levels of 1 μg and 0.001 μg for the reference compounds amphotericin B and miconazole, respectively. No activity against *Cladosporium cucumerinum* was observed (Rath *et al.*, 1996).

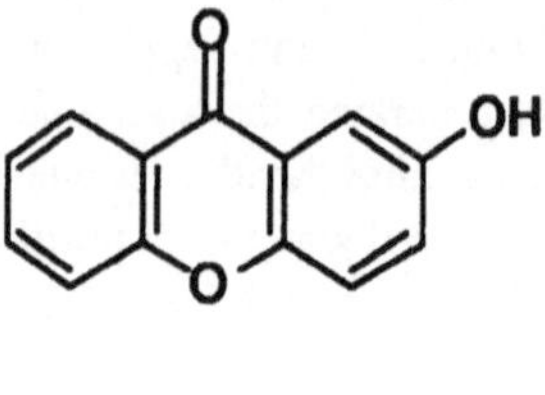
OH
O
O

3

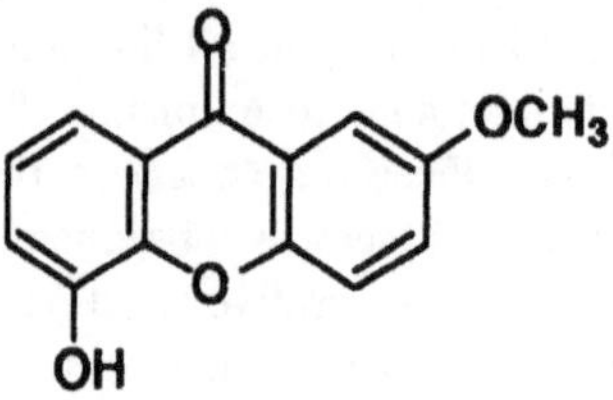
OCH₃
OH
O
O

4

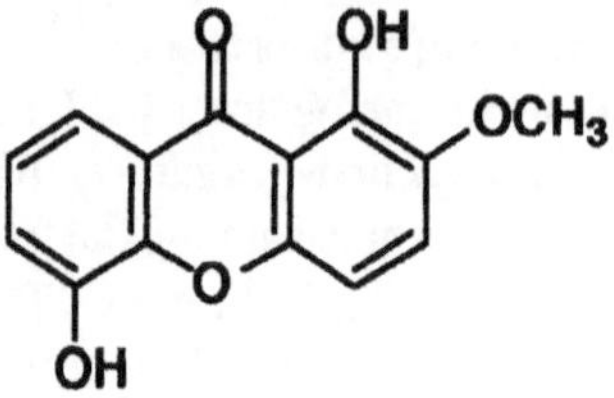
OH
OCH₃
OH
O
O

5

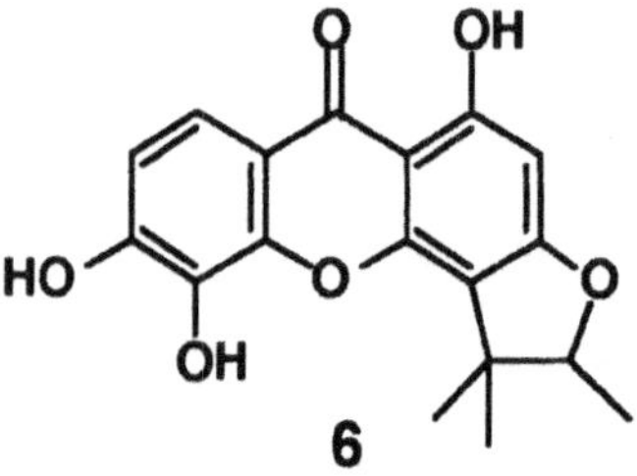
O
OH
HO
OH
O
O

6

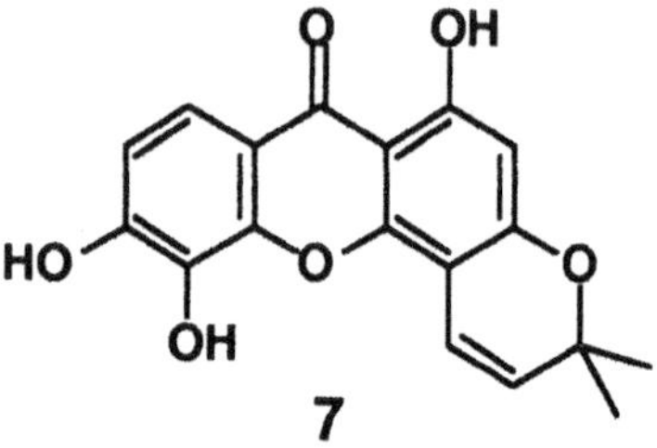
O
OH
HO
OH
O
O

7

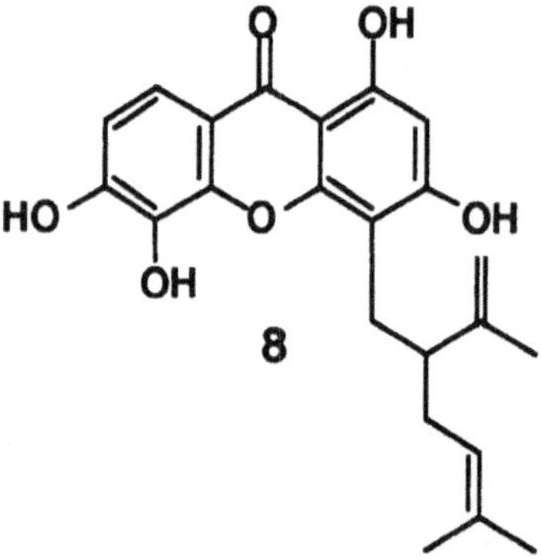
O
OH
HO
OH
OH
O

8

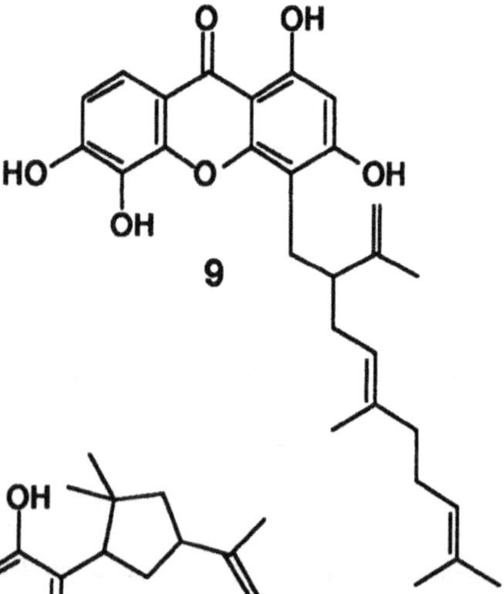
O
OH
HO
OH
OH
O

9

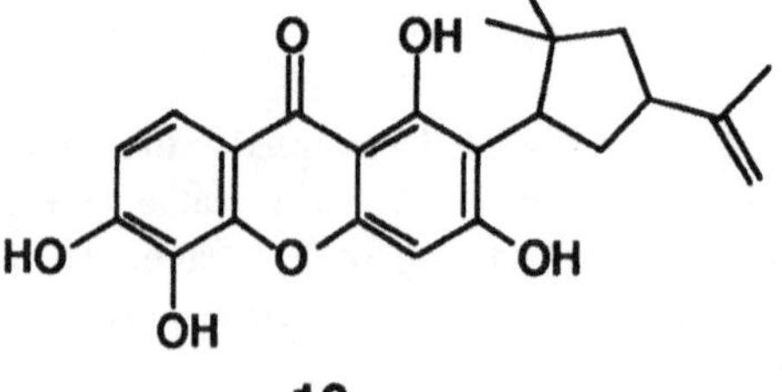
O
OH
HO
OH
OH
O

10

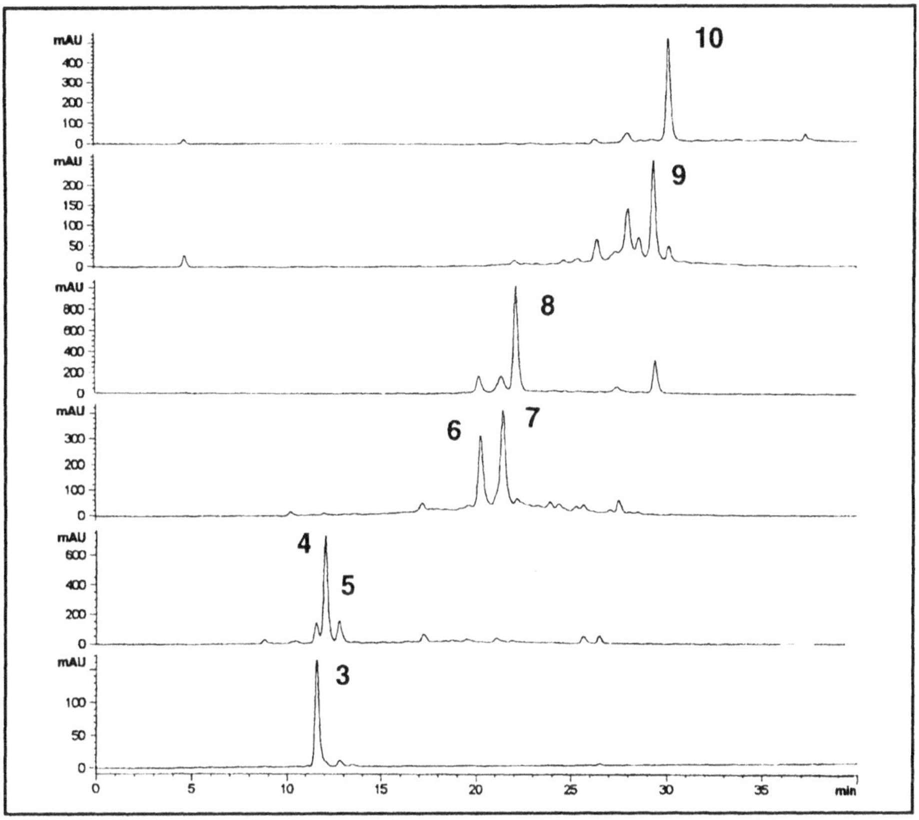

Figure 7. Analytical HPLC of the fractions from the CPC separation of *H. roeperanum* xanthones.
HPLC: see Figure 6.
CPC conditions: Pharma-Tech CCC-1000 instrument; solvent hexane-ethyl acetate-methanol-water 1:1:1:1; upper phase as mobile phase; sample size 363 mg.

3.4. IRIDOID GLUCOSIDES WITH RADICAL SCAVENGING PROPERTIES FROM *FAGRAEA BLUMEI* (LOGANIACEAE)

Fagraea blumei (Loganiaceae) is a tree growing in southeast Asia. As part of our search for new antioxidants in higher plants, we detected a series of radical scavengers in the methanolic stem bark extract of *F. blumei* . These compounds exhibited on TLC plates a strong yellow or blue fluorescence which prompted us to undertake their isolation. The stem bark was successively extracted with dichloromethane and methanol. The methanolic extract was then fractionated by Sephadex LH-20 followed by centrifugal partition chromatography, medium pressure liquid chromatography and high performance liquid chromatography on RP-18 to afford blumeosides A-D (**11-14**) (Cuendet *et al.*, 1997).

Blumeosides are new iridoid glucosides containing either a hydroxy or a dihydroxy terephthalyl moiety. While iridoids and secoiridoids are widespread in Loganiaceae, terephthalic acid derivatives are rather uncommon in plants.

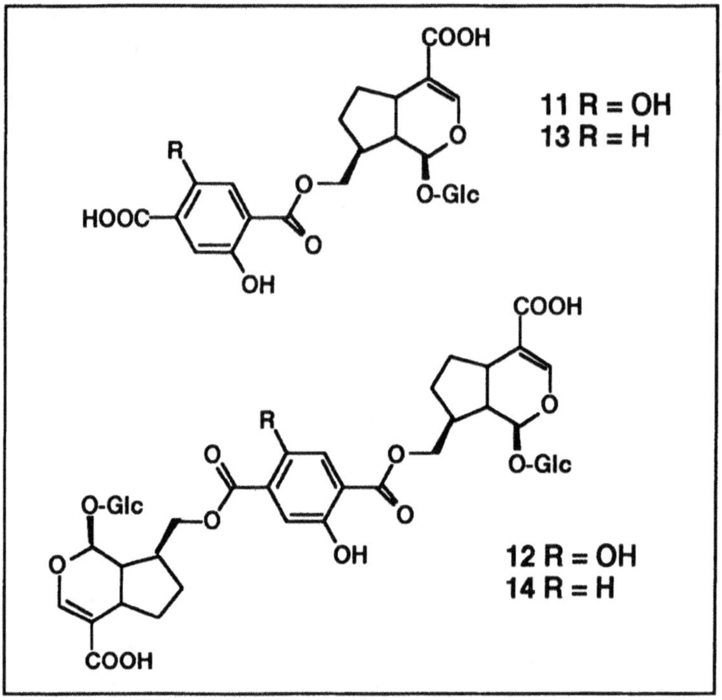

Radical scavenging properties of blumeosides were evaluated against the DPPH radical. By using DPPH as a TLC spray reagent, **11** and **12** (10 mg) appeared as yellow spots against a purple background, while the same amount of **13** and **14** did not react with the radical. Compounds **11-14** were also tested against DPPH in a spectrophotometric assay. Quercetin and BHT were used as reference compounds. The activity of blumeoside A (**11**) remained lower than that of quercetin but was higher than that of BHT. Compound **12** was less active than **11**, while **13** and **14** did not reduce significantly the free radical. Interestingly the two compounds which contain a hydroquinonic moiety exhibited the strongest radical scavenging activity in this assay.

The antioxidative activity of compounds **11-14** was also evaluated spectrophotometrically on the bleaching of the water soluble carotenoid crocin. Compounds **11-14** were all active in this assay but their potency remained lower than that of rutin. Among the iridoids, **11** and **14** exhibited the strongest activity, comparable to that of gallic acid.

4. Conclusions

Plants contain thousands of constituents and are a valuable source of new and biologically active molecules. For their investigation, it is important to have the necessary tools at hand. These include suitable biological assays and chemical screening methods.

Bioassays should be as simple as possible and attempts should be made to have access to a large number of different tests so that many biological properties can be screened. Existing assays, however, are often not reliably predictive for clinical efficiency and care should be taken when interpreting the results.

The bioassays summarized here involve antifungal, antibacterial and antioxidant/radical scavenging activities. They are most effective when used in conjunction with chemical screening methods so that ubiquitous and unimportant compounds can be excluded. For chemical screening, HPLC coupled with different detection methods e.g. UV, MS provides a great deal of preliminary information about the content and nature of constituents found in the active extracts. In certain cases, combination with a spectral library and pre- or post-column derivatization allows structure determination on-line. By selective ion monitoring in LC/MS or even LC/MSMS, it is possible to achieve the detection of specific target molecules - those, for example, which have already been found to exhibit a particular activity. The recent introduction of other hyphenated techniques such as LC/NMR (Spraul *et al.*, 1993) will render the on-line structure determination of metabolites even more accurate and rapid.

Once the novelty or utility of a given constituent is established, it is then necessary to process the plant extract in the usual manner, to isolate samples for full structure elucidation and biological testing.

The combination of biological and chemical screening provides important information about plant constituents but will not be a sufficient condition for the discovery of potent new drugs if suitable pharmacological models (or disease-specific assays in agrochemistry) are not available. It is thus essential to adopt a multidisciplinary approach when working in this field. Efficient collaborations with pharmacologists and medical doctors, plant pathologists and biologists is crucial to see the complete development of an interesting lead compound into a exploitable product.

References

Albert, K. (1995) On-line use of NMR detection in separation chemistry, *J. Chromatogr. A* **703**, 123-147.

Cuendet, M., Hostettmann, K., Potterat, O., and Dyatmiko, W. (1997) Iridoid glucosides with free radical scavenging properties from *Fagraea blumei*, *Helv. Chim. Acta* **80**,

Farnsworth, N.R., Akerele, O, Bingel, A.S., Soejarto, D.D., and Guo, Z.G. (1985) *Medicinal plants in therapy*, Bull. W.H.O. **63**, 965-981.

Hamburger, M.O., and Cordell, G.A. (1987) A direct bioautographic TLC assay for compounds possessing antibacterial activity, *J. Nat. Prod.* **50**, 19-22.

Homans, A.L., and Fuchs, A. (1970) Direct bioautography on thin-layer chromatograms as a method for detecting fungitoxic substances, *J. Chromatogr.* **51**, 325-327.

Hostettmann, K., and Marston, A. (1990) Bioactive constituents of plants used in African traditional medicine, in Atta-ur-Rahman (ed.), *Studies in Natural Products Chemistry, Vol. 7*, Elsevier, Amsterdam, pp. 405-437.

Hostettmann, K., Domon, B., Schaufelberger, D., and Hostettmann, M. (1984) On-line high-performance liquid chromatography: ultraviolet-visible spectroscopy of phenolic compounds in plant extracts using post-column derivatization, *J. Chromatogr.* **283**, 137-147.

Hostettmann, K., Marston, A., and Wolfender J.-L. (1995) Strategy in the search for new biologically active plant constituents, in K. Hostettmann, A. Marston, Maillard and M. Hamburger (eds.), *Phytochemistry of Plants Used in Traditional Medicine*, Clarendon Press, Oxford, pp. 17-45.

Hostettmann, K., Wolfender, J.-L., and Rodriguez, S. (1997a) Rapid detection and subsequent isolation of bioactive constituents of crude plant extracts, *Planta Med.* **63**, 2-10.

Hostettmann, K., Terreaux, C., Marston, A., and Potterat, O. (1997b) The role of planar chromatography in the rapid screening and isolation of bioactive compounds from medicinal plants, *J. Planar Chromatogr.* **10**, 251-257.

Marston, A., and Hostettmann, K. (1994) Counter-current chromatography as a preparative tool - applications and perspectives, *J. Chromatogr. A* **658**, 315-341.

Marston, A., Maillard, M., and Hostettmann, K. (1997) The role of TLC in the investigations of medicinal plants of Africa, South America and other tropical regions, *GIT Laboratory Journal* No. 1, 36-39.

O'Neill, M.J. and Lewis, J.A. (1993) The renaissance of plant research in the pharmaceutical industry, in A.D. Kinghorn and M.F. Balandrin (eds.), *Human Medicinal Agents from Plants*, American Chemical Society, Washington, DC, pp. 48-55.

Rahalison, L., Hamburger, M., Hostettmann, K., Monod, M., and Frenk, E. (1991) A bioautographic agar overlay method for the detection of antifungal compounds from higher plants, *Phytochem. Anal.* **2**, 199-203.

Rath, G., Potterat, O., Mavi, S., and Hostettmann, K. (1996) Xanthones from *Hypericum roeperanum*, *Phytochemistry* **43**, 513-520.

Rios, J.L., Recio, M.C., and Villar, A. (1988) Screening methods for natural products with antimicrobial activity: a review of the literature, *J. Ethnopharmacol.* **23**, 127-149.

Rodriguez, S., Wolfender, J.-L., Hakizamungu, E., and Hostettmann, K. (1995) An antifungal naphthoquinone, xanthones and secoiridoids from *Swertia calycina*, *Planta Med.* **61**, 362-364.

Saxena, G., Farmer, S., Towers, G.H.N., and Hancock, R.E.W. (1995) Use of specific dyes in the detection of antimicrobial compounds from crude plant extracts using a thin layer chromatography agar overlay technique, *Phytochem. Anal.* **6**, 125-129.

Spraul, M., Hoffmann, M., Lindon, J.C., Nicholson, J.K., and Wilson, I.D. (1993) Liquid chromatography coupled with high field proton nuclear magnetic resonance spectroscopy: Current status and future prospects, *Analytical Proceedings* **30**,

Sticher, O. (1993) Quality of *Ginkgo* preparations, *Planta Med.* **59**, 2-11.

Terreaux, C., Maillard, M., Hostettmann, K., Lodi, G., and Hakizamungu, E. (1994) Analysis of the fungicidal constituents from the bark of *Ocotea usambarensis* Engl. (Lauraceae), *Phytochem. Anal.* **5**, 233-238.

Watt, J.M., and Breyer-Brandwijk, M.G. (1962) *The Medicinal and Poisonous Plants of outhern and Eastern Africa*, E. and S. Livingstone, Edinburgh.

Wolfender, J.-L., and Hostettmann, K. (1995) Applications of liquid chromatography - mass spectrometry in the investigation of medicinal plants, in J.T. Arnason, R. Mata, Romeo (eds.), *Phytochemistry of Medicinal Plants*, Plenum Press, New York, 189-215.

NATURAL PRODUCTS WITH ANTIPROTOZOAL ACTIVITY

S. L. CROFT and C. R. WEISS
London School of Hygiene and Tropical Medicine, Department of Infectious and Tropical Diseases, Keppel Street, London WC1E 7HT, United Kingdom

1. Introduction

Protozoal diseases are a major threat to the health of human populations and domestic animals worldwide, mainly due of the absence of vaccines and deficiencies in chemotherapy and vector control. Although chemotherapy has played a major role in the treatment and control of protozoal diseases since the pioneering work of Ehrlich at the beginning of the twentieth century, there remains a wide variation in efficacy of available drugs. Several major advances in antiprotozoal chemotherapy have been achieved in this century, some following the development of synthetic drugs for example the quinolines, diaminopyrimidines and triazenes for malaria and toxoplasmosis, organometallic drugs and diamidines for trypanosomiasis and leishmaniasis, 5nitroimidazoles for amoebiasis, giardiasis and trichomoniasis and hydroxy-naphthoquinones for theileriosis and malaria (Croft, 1997). Natural products have also made an important impact. The polyene antibiotic ionophores monensin, lasalocid, narasin and salinomycin, are widely used in the prophylaxis of avian coccidiosis, and amphotericin B, is given in the treatment of leishmaniasis and some forms of amoebiasis. The aminoglycoside-aminocyclitol paromomycin (aminosidine) is increasingly used in leishmaniasis while tetracyclines and clindamycin are part of the armoury for malaria. Quinine, identified as the active ingredient of Cinchona bark in 1820, is still used extensively in malaria therapy, and artemisinin (Qinghaosu), an active compound of *Artemisia annua* (Asteraceae), is considered to be the most promissing lead amongst the new antimalarial drugs. Emetine, isolated from the rhizome and root of *Cephaelis ipecacuanha* (Rubiaceae) in 1828 was for many years used in the treatment of amoebiasis, although now replaced by it´s derivative dehydroemetine.

The most profound effects of the deficiencies in antiprotozoal chemotherapy are observed in tropical and developing countries. Malaria, a disease causing 250 million infections and over 2 million deaths per annum, regained its position as the most important human infectious disease following the development of resistance by *Plasmodium falciparum* to chloroquine, as well as to pyrimethamine, mefloquine and quinine (Wernsdorfer, 1994). The therapies for human African trypanosomiasis (sleeping sickness), South American trypanosomiasis (Chagas´ disease) and leishmaniasis, diseases caused by closely related haemoflagellate parasites, rely upon arsenical, nitroheterocyclic and antimonial drugs, respectively. These drugs have toxic side effects, variable efficacy and require long courses of treatment (Croft *et al.*, 1997a). Opportunistic protozoan parasites, some unknown in humans only a decade ago, have become major causes of mortality and morbidity in immunocompromised humans. In this category there are no drugs for the treatment of cryptosporidiosis and some forms of microsporidiosis.

The empirical and traditional use of plant products in the 19th century, for example of quinine and emetine, indicated the potential of natural products in disease therapy. However, despite extensive studies to find antiprotozoal activity in plant extracts and isolated compounds (Phillipson & Wright, 1991; Wright & Phillipson, 1990), few plant derived drugs have been developed this century for antiprotozoal chemotherapy. This chapter will focus on the role of plant products as antiprotozoal agents and examine directions and bioassay models used in the search for new drugs and lead compounds.

2. Malaria

CURRENT SITUATION

Four species of *Plasmodium*, *P. falciparum*, *P. malariae*, *P. ovale* and *P. vivax*, cause malaria in humans. P. falciparum is the most lethal and widespread threat to human populations in Africa, South East Asia and South America. All species have a first developmental stage in hepatocytes from which merozoite forms are released to invade erythrocytes and to begin a series of division (schizogony) and re-invasion cycles. Malaria has emerged as a public health problem over the past three decades due to the development of world-wide resistance to chloroquine, a cheap, oral drug with therapeutic and prophylactic activity against the species that infect the humans. Antimalarial drugs developed to replace chloroquine, mefloquine (a 4quinoline-methanol) and halofantrine (a 9-phenanthrenemethanol) were introduced in the 1980s. Unfortunately, they have proved to be less acceptable to patients and resistance to both has appeared in *P. falciparum* (Wernsdorfer, 1994). The use of antifolate drugs, pyrimethamine and cycloguanil, is also compromised by resistance (Wernsdorfer, 1994). Some other drugs, atovaquone (a hydroxynaphthoquinone), pyronaridine (an azacrine) and benflumetol (a fluoromethanol) are on clinical trial (Olliaro & Trigg, 1995). The sesquiterpene lactones, for example artemisinin (see below), represent a new class of antimalarials, which offer considerable hope for the therapy of *P. falciparum* malaria.

PLANT PRODUCTS

Although many plant derived compounds do continue to make a significant impact on antimalarial chemotherapy, it is not true that most commonly used antimalarials are based upon plant-derived lead compounds. The most well known plant derived drug is the quinoline alkaloid quinine, isolated from the bark of south American *Cinchona* species (Rubiaceae). The claimed traditional antimalarial use for Cinchona bark seems to be unlikely as evidence suggests that malaria was introduced to South America by colonisers (Warhurst, 1997). Whatever the origin of use, quinine has become an important alternative treatment in areas where chloroquine resistance in *P. falciparum* predominates (Warhurst, 1997). Unfortunately, resistance to quinine, first reported from South America in 1910, has now also been reported in Africa and South East Asia (Wernsdorfer, 1994). The stereoisomers of quinine, as well as semi-synthetic and synthetic derivatives of quinine, have activity against a number of species of *Plasmodium* (Hofheinz & Merkli, 1984). Quinidine has been used as an antimalarial alone and in combination with quinine, quinidine and cinchonine (Rogier *et al.*, 1996). Quinine formed the basis for the design of mefloquine *(figure 1)* following observations that the metabolic oxidation of the parent alkaloid could be blocked through the introduction of substituents to the nucleus (Sweeney, 1994). However, the aniline dye methylene blue, observed by Guttman and Ehrlich in 1891 to have activity against malaria parasites in patients, was the basis for the synthesis of the acridines and 4- and 8-aminoquinolines such as mepacrine, chloroquine and primaquine (Greenwood, 1992).

The identification of the sesquiterpene lactone, artemisinin as the antimalarial principal isolated from *Artemisia annua* L. (Asteraceae), a plant used in traditional Chinese medicine for over two millennia, has been the most significant discovery for antimalarial chemotherapy in the past 50 years. Since initial observations by Chinese scientists in 1972, artemisinin and its derivatives have been used to treat over 3 million cases of falciparum malaria in South East Asia (White, 1994). More soluble derivatives of artemisinin with improved pharmacokinetics, artemether, artesunate and artether, are now being produced by pharmaceutical companies outside China for oral, parenteral and suppository administration (Olliaro & Trigg, 1995). Artemisinin contains an unusual endoperoxide linkage which appears to be responsible for its antimalarial activity (Meshnick *et al.*, 1996). This observation provided the stimulus for the synthesis of novel trioxane and dioxane molecules *(figure 1)* which have shown promising activity in experimental models (Meshnick *et al.*, 1996). A Yingzhaosu A derivatised sesquiterpene peroxide, arteflene (Ro 42-1611), a product of the plant *Atrabotrys uncinatus* (Annonaceae), reached clinical trials but was discontinued due to high recrudescence rates (Radloff *et al.*, 1996). Plant derived napthoquinones also provided a novel lead for antimalarials. Following the identification of the antimalarial activity of hydrolapachol in the 1940s many analogues were synthesised, including lapinone (a 1,4-hydroxynaphthoquinone, HNQ) and later, in the 1960s, menoctone; both reached clinical trials (Hudson, 1984). Further development of HNQs, culminated in atovaquone *(figure 1)*, which in combination with proguanil is on trial for the treatment of falciparum malaria (Hudson, 1993; Olliaro & Trigg, 1995).

The search for new antimalarial compounds from natural products is a subject of much current research, involving both plants used in traditional medicine and those considered to be a potential source of novel chemical structures. Many compounds isolated from plants have shown strong antiplasmodial activity *in vitro* and represent potential leads for new antimalarial drugs. These studies have been reviewed extensively and the importance of chemical groups well defined (Kirby, 1996; Phillipson & Wright, 1991; Phillipson *et al.*, 1995; Wright & Phillipson, 1990). Unfortunately, few have shown profound activity *in vivo* models of infection. Rational approaches are also being used in the search for new drugs, including structure based modelling of characterised and validated enzyme targets, for example dihydrofolate reductase and proteases. Synthetic derivatives of chalcones, modelled as inhibitors of cysteine proteases and which are essential enzymes in the digestive processes of the malaria parasite, have been shown to kill *P. falciparum in vitro* (Li *et al.*, 1995). Other biochemical approaches offer further opportunities: the methoxylated flavones, artemin and casticin, known to inhibit influx of L-glutamine and myoinositol across the membranes of *P. falciparum* infected erythrocytes, act synergistically with artemisinin (Elford *et al.*, 1987).

BIOASSAYS

Prior to the culture of the asexual cycle of *P. falciparum* in human erythrocytes (Jensen & Trager, 1977) drug screening was based upon *in vivo* avian and rodent models (see Peters, 1987). Since 1977 in vitro assays for measuring drug sensitivity of intraerythrocytic asexual forms of *P. falciparum* have been developed. A semiautomated microdilution assay was described by Desjardins *et al.* (1979) in which inhibition of uptake of a nucleic acid precursor, ^{3}[H]-hypoxanthine, into the parasite is a measure of drug activity. The technique has been modified several times (Geary *et al.*, 1983; O'Neill *et al.*, 1985) but essentially remains the most widely used assay. However, the technique gives limited information on parasite growth or multiplication, and inhibition of precursor uptake may not always related to antimalarial activity. Modifications have been made to improve the sensitivity to antifolates (Tan-Ariya *et al.*, 1987) and to study differential effects of drugs on trophozoite and schizont stages (Ter Kuile *et al.*, 1993). Serum free media have also been described which could provide less expensive assays (Ofulla *et al.*, 1993).

Biochemical and molecular techniques have also been utilised to measure *in vitro* drug activities. Basco *et al.* (1995) described a colorimetric enzyme assay based on the measurement of the lactate dehydrogenase activity in living *P. falciparum* and *P. ovale*. Lactate dehydrogenase is the terminal enzyme of anaerobic glycolysis and uses 3-aceylpyridine-NAD (APAD) as a coenzyme to form pyruvate from L-lactate. This reaction results in the formation of reduced APAD which can in turn reduce nitro blue tetrazolium, forming a formazan product detectable by spectrophotometry.

Flow cytometry was used to study parasitaemia, development of parasites, drug activity and resistance (Van Vianen *et al.*, 1990). The major disadvantages of flow cytometry are a lack of automation for handling many samples and the production of semi-quantitative data. DNA and RNA amplification techniques, such as the Polymerase Chain Reaction (PCR) and Reverse Transcriptase Polymerase Chain Reaction (RT-PCR), largely used in diagnostics, have not yet been adapted for antimalarial drug assays. 96-well microtitre plates adapted assays using DNA probes conjugated with alkaline phosphatase (McLaughlin *et al.*, 1988) or biotin could be modified for a colorimetric assay. As DNA hybridisation

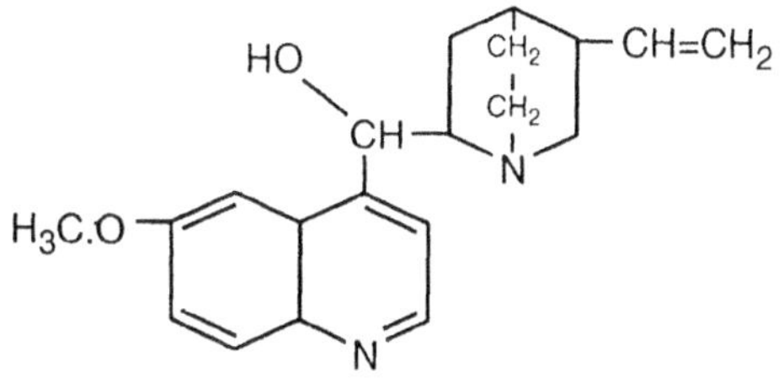

quinine

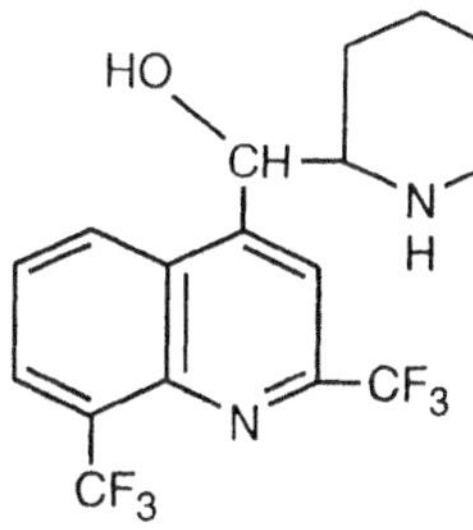

mefloquine

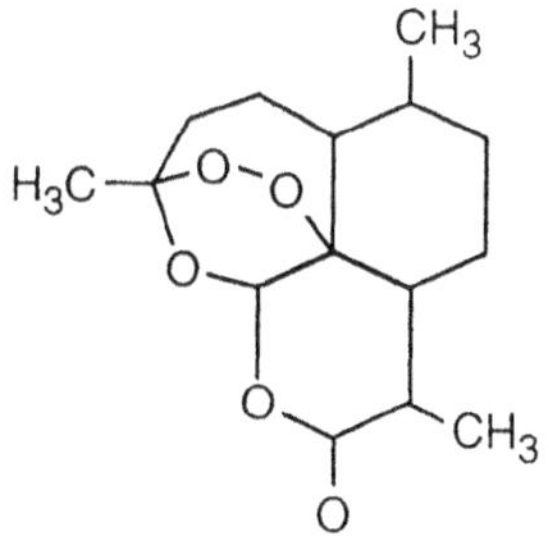

artemisinin

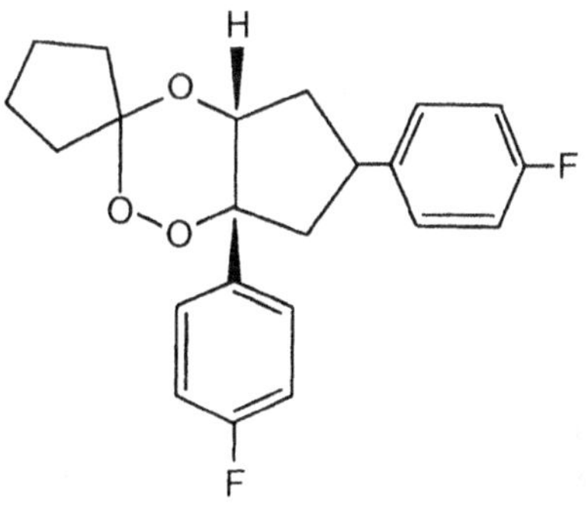

Fenozan-50-F

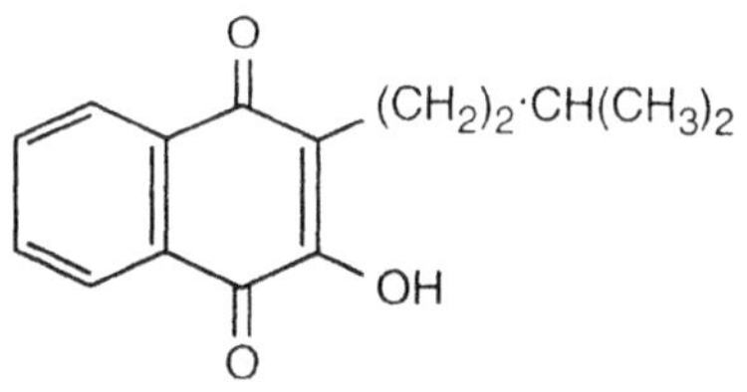

hydrolapachol

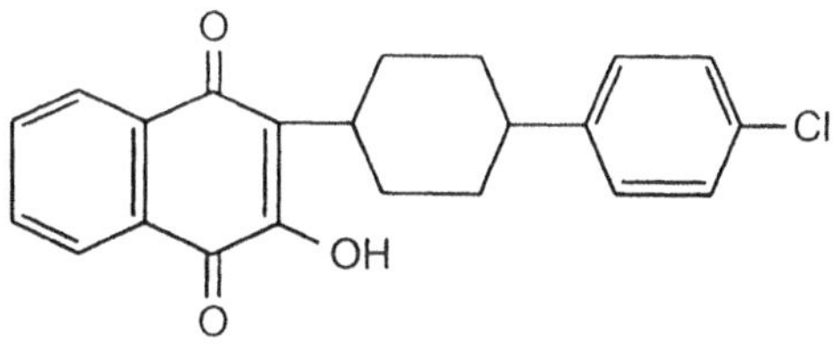

atovaquone

Figure 1. **Plant products and synthetic derivatives with activity against *Plasmodium spp.***

techniques may detect residual DNA of dead parasites the measurement of ribosomal RNA through RT-PCR is a more reliable indication of viability.

Drug sensitivity testing for *P. vivax* and *P. ovale* is a problem as continuous culture of these species is difficult, probably due to the different nutritional and host cell requirements. Only short term *in vitro* cultures of *P. vivax* and *P. ovale* have been used in drug susceptibility testing (Basco *et al.*, 1994). *In vitro* culture of the exoerythrocytic liver stages of malaria infection has been achieved using hepatocytes and hepatomas to culture *P. berghei, P. vivax* and *P. falciparum* (Janse & Waters, 1995; Sinden *et al.*, 1990). The models were used for the *in vitro* evaluation of tissue schizonticidal activity of atovaquone (Davies *et al.*,1993). Li *et al.* (1991) used an rRNA probe to quantitate exoerythrocytic development of *P. berghei* in mouse hepatocyte cultures as well as to evaluate in vitro tissue schizonticidal activity and hepatotoxicity of antimalarial drugs. One major limitation of *in vitro* studies is that isolates which establish easily in culture are not necessarily representative of those being transmitted in human populations. Other *in vitro* tests are available to assess the sensitivity of circulating parasites to antimalarial drugs by microscopical measurement of the inhibition of the maturation of ring forms to schizonts after incubation of parasitized blood at a range of drug concentrations for a period of 24-30 hours′ activity. The WHO Microtest II, requiring just a capillary tube of blood, is widely used in field laboratories to evaluate a patients′ response to schizonticidal drugs (WHO, 1984).

Rodent models of *P. berghei* and *P. yoelii* infections remain an integral part of drug development for detecting blood schizonticidal, tissue schizonticidal and repository activity. The development of well-characterised drug-resistant strains of these parasites has added to their use in identification of novel drugs and drug combinations (Peters, 1987). However, rodent malaria models have been criticised as the biology of these species differs significantly from that of *P. falciparum* and *P.vivax*. As human malaria species do not infect rodents, primate models have been established. *Aotus* and *Saimiri* monkeys have been used in drug studies but ethical issues and cost limit their use. SCID mice have been investigated as alternative hosts both for study of pre-erythrocytic and erythrocytic stages of *P. falciparum* (Badell *et al.*, 1995) but do not appear to be suitable for drug studies.

3. Leishmaniasis

CURRENT SITUATION

Leishmaniasis is a complex of diseases spread throughout the tropical and subtropical world caused by haemoflagellate protozoan parasites. There are 15 species of *Leishmania* that infect humans, some, for example *L. donovani*, causing the potentially fatal visceral disease, and others, for example *L. major* and *L. braziliensis* causing disfiguring cutaneous disease. In humans, and other mammalian hosts, *Leishmania* are obligate parasites of macrophages, surviving and multiplying in the phagolysosomal compartment of these cells. Pentavalent antimonials, sodium stibogluconate and meglumine antimonate, have been the first line drugs for the treatment of both visceral leishmaniasis (VL) and cutaneous leishmaniasis (CL) for over 50 years. They require long courses with parenteral administration, have toxic side effects and variable efficacy. Increasing resistance to antimonials is being observed in VL (Olliaro & Bryceson, 1993). Alternative drugs include the polyene antibiotic amphotericin B, which is also available in highly effective, less toxic but expensive lipid formulations (Davidson *et al.*, 1996), and the aminoglycoside paromomycin (aminosidine) used in a parenteral formulation for VL and a topical formulation for CL (Olliaro & Bryceson, 1993).

PLANT PRODUCTS

Plant extracts have been used traditionally in the treatment of leishmaniasis, in particular CL. There is a considerable literature on the use of the alkaloid berberine for CL, including the first recorded clinical trial in 1911, and studies on intralesional application. Vennerstrom *et al.* (1990) examined the structure-activity relationship of berberine derivatives against experimental CL. Emetine, it's derivatives and other alkaloids have been tested against experimental leishmaniasis (Neal, 1987) and more recently a series of 2-substituted quinoline alkaloids (chimanines) have shown promising activity *in vivo* (Fournet *et al.*, 1996). Other compounds for example licochalcone A, isolated from Chinese liquorice roots (Chen *et al.*, 1994) and naphthoquinones, for example plumbagin (Croft *et al.*, 1985; Fournet *et al.*, 1992) have also shown experimental *in vivo* activity *(figure 2)*. Activities of many other plant products have been reviewed by Iwu *et al.* (1994).

BIOASSAYS

The extracellular promastigote is easily grown in culture and was used in simple assays. *Leishmania* promastigote viability has been determined isotopically or by vital stains or dyes (Croft, 1986). Promastigote assays have been modified to improve drug sensitivity (Jackson *et al.*, 1989) and in order to compare the sensitivity of clinical isolates to standard antimonials (Grogl *et al.*, 1992). However, it is the intracellular amastigote that is relevant for chemotherapy and this form has a different biochemistry and drug sensitivity to the promastigote. Compounds therefore should be tested *in vitro* against intracellular amastigotes in macrophage models which can sustain a multiplying population of amastigotes, can tolerate a wide range of drug concentrations and be infected by different *Leishmania* species which exhibit widely different drug susceptibilities (Navin *et al.*, 1992, Allen & Neal, 1995). In the frequently used models, rodent macrophage cell lines, primary isolated mouse peritoneal or human monocyte-derived infected macrophages are maintained in a drug containing medium for 4 to 7 days. Drug activity can be assessed by such various methods as (i) microscopical counting of stained preparations (Neal & Croft, 1984), (ii) incorporation of radiolabelled ^{3}[H]-uracil into amastigotes (Berman & Gallalee, 1985) or (iii) determining the number of promastigotes which transform from surviving amastigotes. New methods are required to determine the number of surviving amastigotes in drug treated

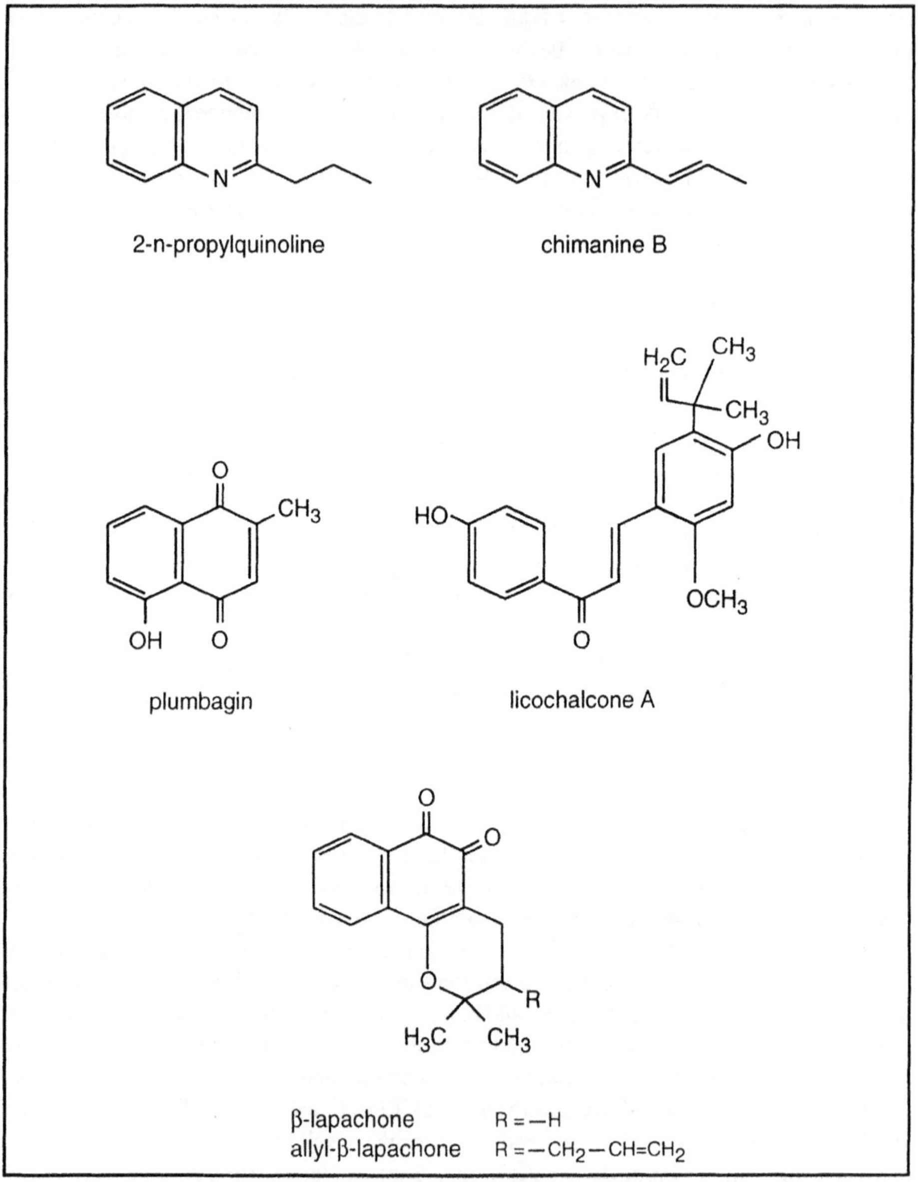

Figure 2. Plant products and derivatives with activity against *Leishmania spp*. and *Trypanosoma cruzi*.

cultures. Simple and quick PCR methods have been used to detect parasites in macrophages (Nuzum *et al.*, 1995). Axenic cultures of *Leishmania* amastigotes have been sucessfully established which enable direct evaluation of the activity of antileishmanial compounds on amastigotes independently of the host macrophage (Ephros *et al.*, 1997). Axenic cultures have potential for large primary screening assays but do not take account of the effects of host cell metabolism, drug accumulation or intracellular pH on drug activity.

Until 20 years ago most of the drug screening was conducted in rodent models. Currently, the most frequently used *in vivo* models for VL are BALB/c mice and hamsters. Rodents are infected intravenously by promastigotes or amastigotes and the drugs are tested against established infections. Drug activity is determined by counting the number of amastigotes in liver and spleen. CL infections, especially of *L. major* and *L. mexicana* are frequently used as well as the more difficult but nevertheless important models with *L. braziliensis* and *L. tropica*. BALB/c mice are anergic to *L. major* and this non cure model is a demanding test for any drug. Self cure models in other strains of mice are also available. Drug activity in CL is normally determined by the measurement of the change of lesion size, but methods to quantitate the number of surviving amastigotes in lesions have been described (Lima *et al.*, 1997).

4. South American Trypanosomiasis

CURRENT SITUATION

In Latin America there are up to 18 million cases of South American trypanosomiasis (Chagas´ disease) caused by the haemoflagellate parasite *Trypanosoma cruzi*. The parasite has a non-dividing invasive trypomastigote stage in the blood (which can be transmitted during blood transfusion) and amastigotes which survive and multiply in the cytoplasm of macrophages, muscle and nerve cells. Two nitroheterocyclic compounds introduced in the 1970s, benznidazole and nifurtimox, are the recomended drugs for the treatment of Chagas´ disease. Both drugs have toxic side effects, limited efficacy during the chronic phase of the disease and variable efficacy against different strains of *T. cruzi*. The purine analogue allopurinol and the sterol biosynthesis inhibitors, ketoconazole and itraconazole, had limited sucess in clinical trials (Croft *et al.*, 1997a).

PLANT PRODUCTS

There is no obvious traditional medicine for Chagas´ disease, probably due to the insidious nature of this infection. However, in the search for new drugs considerable efforts have been made to identify new compounds from plants in South America and elsewhere and the mechanisms of action of several active plant products have been studied (see De Castro, 1993; Sepulveda-Boza & Cassels, 1996). The sensitivity of *T. cruzi* to radicals is one approach that has been exploited: redox cycling quinones (naphthoquinones, benzoquinones, mansonones) have been shown to generate cytotoxic oxygen radicals which can kill *T. cruzi* (Docampo & Moreno, 1984). The β-lapachone derivative, allyl-β-lapachone (*figure 2*), showed some of the highest activity against trypomastigotes and amastigotes *in vitro*, but poor *in vivo* efficacy (Lopes *et al.*, 1978). Taxol, a diterpene alkaloid and anticancer drug which works through microtuble stabilisation, inhibited the growth of epimastigotes at 10μM (Baum *et al.*, 1981). The garlic derived organosulphur ajoene, was active *in vitro* against both extra- and intracellular stages through the inhibition of phosphatidylcholine biosynthesis (Urbina *et al.*, 1993).

In trypanosomes and leishmanias the major thiol involved in protection against toxic metabolites and maintenance of intracellular redox balance is the dithiol trypanothione (bis-glutathionyl-spermidine). Trypanothione is maintained in a reduced state by trypanothione reductase (TR), an enzyme with different substrate specificity to glutathione reductase in mammalian cells. TR is therefore a rational chemotherapeutic target and naphthoquinone and nitrofuran derivatives have been designed which act as subversive substrates (Henderson *et al.*, 1988). Expressed TR from *T. cruzi* has been used in automated screens. The cysteine protease cruzipain, has been studied as a rational drug target and the expressed enzyme also has been used in automated screens.

BIOASSAYS

The chemotherapeutic targets in Chagas´ disease are (i) the invasive non-dividing bloodstream trypomastigotes and (ii) the dividing intracellular amastigotes. *In vitro* models involving both stages of the *T. cruzi* life cycle are therefore important. Drug tests use primarily the dividing intracellular stage in different host cells types, for example myoblasts, fibroblasts and macrophages. In this amastigote system infected cells are incubated with drug for 3 to 5 days, as after this time amastigotes transform to trypomastigotes and escape from the host cells. A dividing host cell population, low infection rates and the presence of trypomastigotes in the overlay can complicate the interpretation of these results (Croft *et al.*, 1986). In the trypomastigote test, the trypomastigotes are isolated from infected mouse blood or from *in vitro* culture systems and exposed to the drug for 24h to 48h at either 4°C or 37°C. The test at 4°C is carried out in order to find a suitable drug to be used in blood banks to prevent the transmission of Chagas´ disease during blood transfusion (Croft *et al.*, 1988). In both assays viability and activity is evaluated by microscopical analysis. The recent development of techniques which allow the stable genetic transfection of trypanosomatids make the introduction of large DNA fragments into *T. cruzi* and *L. donovani* by cosmid shuttle vectors possible (Kelly *et al.*, 1994). *T. cruzi* parasites which express the *E. coli*-β-galactosidase gene lacZ (Buckner *et al.*, 1996) have been adapted for use in automated drug screening assay. The β-galactosidase is able to catalyse a colorimetric reaction which can be quantified. As coloured drugs can interfere with the final absorbance step (Buckner *et al.*, 1996) the interpretation of the obtained results with crude plant extracts might cause dificulties. An assay based on genetic transformed parasites is also limited to a primary assay screening because second phase evaluation studies must include a number of isolates of *T. cruzi* from different Latin American regions as significant variation in the sensitivity of strains to standard and experimental drugs is well established (Filardi & Brener, 1987).

The *in vivo* evaluation of drugs against *T. cruzi* also requires improvement. In tests in rodent the determination of the curative activity of a drug involves the detection of parasites in blood and tissues from surviving mice by using haemoculture, PCR, histopathology or sub-inoculation of tissues of treated mice into naive mice.

5. African Trypanosomiasis

CURRENT SITUATION

There are currently up to 200,000 cases/annum of human African trypanosomiasis (sleeping sickness), in West and Central Africa, caused by two closely related parasites *Trypanosoma brucei gambiense* and *Trypanosoma brucei rhodesiense*. These parasites divide rapidly during the early haemolymphatic stage of the disease before penetrating the CNS causing late stage disease. Pentamidine, a diamidine, and suramin, a sulphonated naphthylamine, are used to treat the early haemolymphatic stages, while the trivalent arsenical melarosprol is the standard drug for treating the late stage infection. Eflornithine, an ornithine analogue which inhibits polyamine biosynthesis, was developed as a drug for trypanosomiasis in the 1980s but is only active against late stage *T. b. gambiense* infections. Unfortunately, the requirement for high doses and high costs limit its use (Croft *et al.*, 1997a).

PLANT PRODUCTS

In contrast to the situation with leishmaniasis and Chagas´ disease, where plant products have been identified on the basis of either ethnopharmacology or systematic biochemical and screening approaches, there have been limited studies to identify and evaluate plant products for African trypanosomiasis. Weiss *et al.* (1995) and Freiburghaus *et al.* (1996) have investigated African plants reported as traditional remedies for sleeping sickness; extracts of *Bussea occidentalis*, *Annona senegalensis*, and *Physla angulata* had activity against *T. rhodesiense* in vitro with ED50 values below 5μg/ml and are currently undergoing *in vivo* tests. There have been studies which indicate that *T. brucei* is highly sensitive to quinones but this work involved mainly synthetic compounds (Meshnick *et al.*, 1978). Other examples of limited studies include reports on the *in vivo* activity of extracts of *Annona senegalensis*, Annonaceae (Igweh & Onabanjo, 1989) and biochemical mechanisms of gossypol (Eid *et al.*, 1988), a polyphenolic compound of from the cottonseed of *Gossypium hirsutum* and *G. barbadense*.

BIOASSAYS

Most early *in vitro* screens employed bloodstream trypomastigotes which were isolated from infected rodents and maintained for up to 24 hours in a drug containing medium. The parasites' viability was assessed microscopically, although a later test system used the uptake of 3[H]-hypoxanthine (Desjardins *et al.*, 1980). This test system was unable to detect slow acting trypanocides, used parasites with limited viability and had a short drug incubation time. The first method for continuous cultivation of *T. b. brucei* trypomastigote bloodstream forms, described by Hirumi et al. (1977), used mammalian cell feeder layers. The determination of trypanocidal activities of drugs by measuring inhibition of the growth of dividing bloodstream trypomastigote populations of *T. b. brucei* was described in this model (Borowy *et al.*, 1985; Kaminsky *et al.*, 1987). The presence of a mammalian feeder cell can effect the interpretation of results and limits opportunities for automation. A major advance was the introduction of systems for the axenic cultivation of human pathogenic *T. b. gambiense* and *T. b. rhodesiense* as well as the animal pathogenic *T. evansi, T. congolense, T. vivax* and *T. equiperdum* strains (Baltz *et al.*, 1985; Hirumi & Hirumi, 1989). Drug assays using axenic cultures are now standard. Drug activities can be determined by isotopic (Brun & Kunz, 1989), colorimetric (Zinsstag *et al.*, 1991) or fluorometric (Obexer *et al.*, 1994) assays. To quantify drug activity on slow acting antitrypanosomal drugs such as suramin, a long incubation-low inoculation test (LILIT) where the minimal inhibitory concentration is determined microscopically has been established (Brun & Lun, 1993).

Several rodent models are available to determine the activity of drugs against acute infections of *T. b. brucei* and *T. b. rhodesiense in vivo* but animal models for *T. b. gambiense* are limited. One of the important properties of a drug for the treatment of human African trypanosomiasis is the ability to pass the blood-brain barrier. Mouse and rat models have been used to study CNS infections. In the widely used model of chronic *T. b. brucei* CNS infection the standard drug diminazene is unable to cure mice after day 14-21 of infection when the CNS infection is established. Only drugs which can cross the blood brain barrier are effective against chronic *T. b. brucei* infections (Jennings & Gray, 1983).

6. Opportunistic Protozoa

CURRENT SITUATION

The incidence and severity of human infection by *Toxoplasma gondii, Cryptosporidium parvum, Isospora belli, Leishmania infantum* and Microsporidia have increased dramatically in the past 15 years with the increase in size of the immunocompromised population, in particular those with HIV-infection. The standard antifolate combination of pyrimethamine-sulphadiazine for toxoplasmosis has limited efficacy against cyst stages in the CNS, which are common in AIDS patients. The toxicity and side effects of these drugs preclude their use in 40% of patients. Atovaquone has shown some anti-cyst activity in vitro and in vivo against *T. gondii* and seems to be an equivalent or better alternative to the standard treatment of these infections (Torres *et al.*, 1997). *C. parvum*, hardly known in humans 10 years ago, has been reported in 10-50% of HIV infected patients. There is no approved drug for the treatment of cryptosporidiosis although paromomycin has shown some dose related activity. Eight species of Microsporidia have been detected in humans, several discovered for the first time in AIDS patients during the past 15 years. Normally, immunocompetent patients infected with *L infantum* respond to treatment with antimonials but efficacy of these drugs against the parasite in *L. infantum*-HIV co-infected patients is greatly reduced (Alvar *et al.*, 1997).

BIOASSAYS

There have been limited studies with plant products on any of these parasites because of problems with establishing even basic *in vitro* or *in vivo* models (see below) and immunosuppressed models of infection.

a) Toxoplasma gondii

Toxoplasma gondii is an obligate intracellular parasite which has been grown in a number of different types of fibroblast cells as well as macrophages for *in vitro* drug assays. An assay based upon the incorporation of ^{3}[H]-uracil into the parasite but not the host cell has been the standard method to measure parasite growth in drug-containing medium (Pfefferkorn & Pfefferkorn, 1977). Colorimetric and fluorescent assays have been developed which offer several advantages over radiolabel assays (Derouin & Chastang, 1988) and most recently a system using *T. gondii* expressing β-galactosidase has made high throughput screens possible (McFadden *et al.*, 1997). In infected mammals it is the slow growing form in pseudocysts, muscle and brain tissue which is difficult to treat with medication. *In vitro* models of these tissue pseudocysts have been developed and shown to be useful for drug testing (McHugh *et al.*, 1994). Murine *in vivo* models are well established for testing compounds against acute infections and also against chronic infections with brain cyst stages.

b) Cryptosporidium parvum

It is difficult to establish *C. parvum* infections in rodents. Therefore testsystems have used in chemically immunosuppressed rodent, neonatal or immunodeficent mice (Mead *et al.*, 1995). There is also a lack of a well defined and reliable *in vitro* drug assay for *C. parvum*. Some attempts to design *in vitro* assays have been made where the oocysts are excysted and developed in a mouse fibroblast cell line with drug containing medium (McDonald *et al.*, 1990) or where a cloned, differentiated human enterocyte cell line is used (Marshall & Flanigan, 1992). Both systems determine drug activity against the asexual development of *C. parvum*. Some highly sensitive PCR based detection testsystems of viable parasitic DNA have been recently published (Rochelle *et al.*, 1997). These detection methods could provide a basis for the development of a reliable *in vitro* drug screening assay.

c) Microsporidia
Microsporidia are obligate intracellular parasites that are predominantly associated with AIDS. Eight species have been described causing microsporidosis in humans. There is no tissue culture model for the most common of these, *Enterocytozoon bieneusi*, an important cause of intestinal microsporidiosis (Croft *et al.*, 1997b) In contrast three species of *Encephalitozoon, E. intestinalis, E. hellem* and *E. cuniculi* have been cultured for *in vitro* drug assays and screens are established (Beauvais *et al.*, 1994). Animals models of *Encephalitozoon* are also available for *in vivo* drug test (Didier *et al.*, 1994).

7. Discussion

There is a clear need for new therapeutic and prophylactic drugs for many protozoan diseases of humans and domestic animals. The majority of these diseases are found in developing and tropical regions and consequently there is no economic incentive for the pharmaceutical industries to invest in the identification and development of new antiprotozoal drugs. Therefore it is important to bring together the strengths and expertise of international organisations, academia and the pharmaceutical industries to work on these problems together (Gutteridge, 1997).

Despite the promises of the 19th century, plant products have not contributed as much to the chemotherapy of protozoan diseases in the 20th century as they have to other areas, from antibacterials to immunosuppressive drugs. Apart from the economic factors alluded to above, two further points have been considered to make the identification and evaluation of novel antiprotozoals difficult: (a) the similarity between biochemical pathways in protozoa and mammalian cells, and (b) the complexity of the life cycle of many protozoa. The first point is not readily substantiated as advances in biochemistry and molecular biology have characterised major differences between protozoa and other eukaryotic cells and identified potential drug targets (see Coombs & Croft, 1997; Coombs & North, 1991; Vial, 1996). However, the second point does still cause problems. As already mentioned there are no *in vitro* models for *P. vivax, P. ovale, C. parvum* or some types of Microsporidia. There are no easy *in vivo* models for the malaria parasites of humans, for cryptosporidiosis, for chronic trypanosomiasis, or for some types of leishmaniasis. Other parasite-, host- and drug depend-factors have also to be considered. Drug sensitivity can vary significantly among different species and strains as illustrated by the sensitivity of African trypanosomes to eflornithine, *T. cruzi* to nifurtimox, *Leishmania spp.* to antimonials and azoles and *P. falciparum* strains to most antimalarials. Host dependent factors have assumed much greater importance as some protozoal diseases for example visceral leishmaniasis, toxoplasmosis, cryptosporidiosis and microsporidiosis, have proved to be difficult to treat in immunocompromised humans, especially in HIV coinfections.

The importance of natural products in drug design and strategies involving novel targets, for example cell signalling, has been elegantly described (Verdine, 1996). To exploit the novel developments in chemistry and the availability of natural products it will become essential to: (i) adapt current assays for high throughput screening, more difficult for the important intracellular stages of parasite life cycles, and (ii) to have available expressed enzyme of validated drug targets, for example trypanothione reductase, dihydrofolate reductase/thymidylate synthase, and (iii) to have batteries of species/strains of defined drug sensitivity to screen for novel inhibitors. Molecular biology can also provide methods to improve the rapidity and sensitivity of testing drugs in rodent models of infection.

8. References

Allen S., Neal, R.A. (1989) The *in vitro* susceptibility of macrophages infected with amastigotes of *Leishmania spp.* to pentavalent antimonial drugs and other compounds with special relevance to cutaneous isolates, in D.T. Hart (ed), *Leishmaniasis*, Plenum Publishing Corporation

Alvar, J., Canavate, C., Gutierrez-Solar, B., Jimenez, M., Laguna, F., Lopez-Velez, R., Molina, R., Moreno, J. (1997) *Leishmania* and human immunodeficiency virus coinfection, the first 10 years, *Clinical Microbiology Review*, **10**, 298-319

Badell, E., Pasquetto, V., Eling, W., Thomas, A. (1995) Human *Plasmodium* liver stages in SCDI mice, a feasible model?, *Parasitology Today*, **11**, 169-171

Baltz, T., Baltz, D., Giroud, C., Crockett, J. (1985) Cultivation in a semi-defined medium of animal infective forms of *Trypanosoma brucei, T. equiperdum, T. evansi, T. rhodesiense* and *T. gambiense*, *European Molecular Biology Organisation Journal*, **4**, 1273-1277

Basco, L.K., Le Bras, J. (1994) Short-term *in vitro* culture of *Plasmodium vivax* and *P. ovale* for drug-susceptibility testing, *Parasitology Research*, **80**, 262-264

Basco, L.K., Marquet, F., Makler, M.M., Le Bras, J. (1995) *Plasmodium falciparum* and *Plasmodium vivax*-lactate dehydrogenase activity and its application for *in vitro* drug susceptibility assay, *Experimental Parasitology*, **80**, 260-271

Baum, S.G., Wittner, M., Nadler, J.P., Horwitz, S.B., Dennis, J.E., Schiff, P.B., Tanowitz, H.B. (1981) Taxol, a microtubule stabilizing agent, blocks the replication of *Trypanosoma cruzi*, *Proceedings of the National Academy of Sciences U. S. A*, **78**, 4571-4575

Beauvais, B., Sarfati, C., Challier, S., Derouin, F. (1994) *In vitro* model to assess the effect of antimicrobial agents on *Encephalitozoon cuniculi*, *Antimicrobial Agents and Chemotherapy*, **38**, 2440-2448

Berman, J.D., Gallalee, J.V. (1985) Semiautomated assessment of *in vitro* activity of potential antileishmanial drugs, *Antimicrobial Agents and Chemotherapy*, **28**, 723-726

Borowy, N.K., Hirumi, H., Waithaka, H.K., Mkoji, G. (1985) An assay for screening drugs against animal-infective bloodstream forms of *Trypanosoma brucei brucei in vitro* drugs, *Experimental Clinical Research*, **11**, 155-161

Brun, R., Kunz, C. (1989) *In vitro* drug sensitivity test for *Trypanosoma brucei* subgroup bloodstream trypomastigotes, *Acta Tropica*, **46**, 361-368

Brun,R., Lun, Z.R. (1994) Drug sensitivity of *Chinese Trypanosoma evansi* and *Trypanosoma equiperdum* isolates, *Veterinary Parasitology*, **52**, 37-46

Buckner, F.S., Verlinde, C.L., La-Flamme, A.C., Van Voorhis, W.C. (1996) Efficient technique for screening drugs for activity against *Trypanosoma cruzi* using parasites expressing beta-galactosidase, *Antimicrobial Agents and Chemotherapy*, **40**, 2592-2597

Chen, M., Christensen, S.B., Theander, T.G., Kharazmi, A. (1994) Antileishmanial activity of licochalcone A in mice infected with *Leishmania major* and in hamsters infected with *Leishmania donovani*, *Antimicrobial Agents and Chemotherapy*, **38**, 1339-1344

Coombs G., North M. (1991) The current status of biochemical protozoology, in G. Coombs & M. North (eds), *Biochemical Protozoology*, Taylor & Francis, London, Washington, 1991, pp 23-65

Coombs G.H., Croft S.L. (1997) Molecular basis of drug design and resistance, *Parasitology*, **114** ,Suppl. 1997

Croft S.L. (1997), The current status of antiparasite chemotherapy, *Parasitology*, **114**, 3-15 Suppl. 1997

Croft S.L., Urbina J.A., Brun R. (1997a) Chemotherapy of Human Leishmaniasis and Trypanosomiasis, in *Trypanosomiasis and Leishmaniasis*, eds: Mottram, J.C., Coombs G.H., Holmes, P.H., LAB International, pp: 245-257

Croft S.L., Williams J., McGowan I. (1997b) Intestinal Microsporidiosis, *Seminars in Gastrointestinal Disease*, **8**, 45-55

Croft, S.L. (1986) *In vitro* screens in the experimental chemotherapy of *Leishmaniasis* and *Trypanosomiasis*, *Parasitology Today*, **2**, 64-69

Croft, S.L., Evans, A.T., Neal, R.A. (1985) The activity of plumbagin and other electron carriers against *Leishmania donovani* and *Leishmania mexicana amazonensis*, *Annals of Tropical Medical Parasitology*, **79**, 651-653

Croft, S.L., Walker, J.J., Gutteridge, W.E. (1988) Screening of drugs for rapid activity against *Trypanosoma cruzi* trypomastigotes *in vitro*, *Tropical Medical Parasitology*, **39**, 145-148

Davidson, R.N., di Martino, L., Gradoni, L., Giacchino, R., Gaeta, G.B., Pempinello, R., Scotti, S., Cascio, A., Castagnola, E., Maisto, A., Gramiccia, M., di Caprio, D., Wilkinson, R.J., Bryceson, A.D. (1996) Short-course treatment of visceral leishmaniasis with liposomal amphotericin B (AmBisome), *Clinical and Infectious Diseases*, **22**, 938-943

Davies, C.S., Pudney, M., Nicholas, J.C., Sinden, R.E. (1993) The novel hydroxynaphthoquinone 566C80 inhibits the development of liver stages of *Plasmodium berghei* cultured *in vitro*, *Parasitology*, **106**, 1-6

De Castro, S.L. (1993) The challenge of Chagas' disease chemotherapy, an update of drugs assayed against *Trypanosoma cruzi*, *Acta Tropica*, **53**, 83-98

Derouin, F., Chastang, C. (1988) Enzyme immunoassay to assess effect of antimicrobial agents on *Toxoplasma gondii* in tissue culture, *Antimicrobial Agents and Chemotherapy*, **32**, 303-307

Desjardins, R.E., Canfield, C.J., Haynes J.D., Chulay J.D. (1979) Quantitative assessment of antimalarial activity *in vitro* by a semiautomated microdilution technique, *Antimicrobial Agents Chemotherapy*, **16**, 710-718

Desjardins, R.E., Casero, R.A. Jr., Willet, G.P., Childs, G.E., Canfield, C.J. (1980) *Trypanosoma rhodesiense*, semiautomated microtesting for quantitation of antitrypanosomal activity *in vitro*, *Experimental Parasitology*, **50**, 260-271

Didier, E.S., Varner, P.W., Didier, P.J., Aldras, A.M., Millichamp, N.J., Murphey-Corb, M., Bohm, R., Shadduck, J.A. (1994) Experimental microsporidiosis in immunocompetent and immunodeficient mice and monkeys, *Folia Parasitologica Praha*, 41, 1-11

Docampo, R., Moreno S.N.J. (1984) Free radical metabolites in the mode of action of Chemotherapeutic Agents and phagocytic cells and *Trypanosoma cruzi*, *Reviews of Infectious Diseases*, **6**, 223-238

Eid, J.E., Ueno, H., Wang, C.C., Donelson, J.E. (1988) Gossypol-induced death of african trypanosomes, *Experimental Parasitology*, **66**, 140-142

Elford, B.C., Roberts, M.F., Phillipson, J.D., Wilson, R.J. (1987) Potentiation of the antimalarial activity of qinghaosu by methoxylated flavones, *Transactions of the Royal Society of Tropical Medicine and Hygiene*, **81**, 434-436

Ephros, M., Waldman E., Zilberstein D. (1997) Pentostam induces restistance to antimony and the preservative chlorocresol in *Leishmania donovani* promastigaotes and axenically grown amastigotes, *Antimicrobial Agents and Chemotherapy*, **41**, 1064-1068

Filardi, L.S., Brener, Z. (1987) Susceptibility and natural resistance of *Trypanosoma cruzi* strains to drugs used clinically in Chagas disease, *Transactions of the Royal Society of Tropical Medicine and Hygiene*, **81**, 755-759

Fournet, A., Ferreira, M.E., Rojas,-de Arias, A., Torres,-de Ortiz, S., Fuentes, S., Nakayama, H., Schinini, A., Hocquemiller, R. (1996) *In vitro* efficacy of oral and intralesional administration of 2-substituted quinolines in experimental treatment of new world cutaneous leishmaniasis caused by *Leishmania amazonensis*, *Antimicrobial Agents and Chemotherapy*, **40**, 2447-2451

Fournet,A., Barrios, A.A., Munoz, V., Hocquemiller,R., Cave, A. (1992) Effect of natural naphthoquinones in BALB/c mice infected with *Leishmania amazonensis* and *L. venezuelensis*, *Tropical and Medical Parasitology*, **4**, 219-222

Freiburghaus, F., Kaminsky, R., Nkunya, M.H., Brun, R. (1996) Evaluation of African medicinal plants for their *in vitro* trypanocidal activity, *Journal of Ethnopharmacology*, **55**, 1-11

Geary, T.G., Divo, A.A., Jensen, J.B. (1983) An *in vitro* assay system for the identification of potential antimalarial drugs, *Journal of Parasitology*, **69**, 577-583

Greenwood D. (1992) The quinine connection, *Journal of Antimicrobial Chemotherapy*, **30**, 417-427

Grogl, M., Thomason, T.N., Franke, E.D. (1992) Drug resistance in leishmaniasis: its implication in systemic chemotherapy of cutaneous and mucocutaneous disease, *American Journal of Tropical Medicine and Hygiene*, **47**, 117-126

Henderson, G.B., Ulrich, P., Fairlamb, A.H., Rosenberg, I., Pereira, M., Sela, M., Cerami, A. (1988) "Subversive" substrates for the enzyme trypanothione disulfide reductase: alternative approach to chemotherapy of Chagas disease, *Proceedings of the National Academy of Sciences U. S. A.*, **85**, 5374-5378

Hirumi, H., Doyle, J.J., Hirumi, K. (1977) African trypanosomes, cultivation of animal-infective *Trypanosoma brucei in vitro*.,*Science*, **196**, 992-994

Hirumi, H., Hirumi, K. (1989) Continuous cultivation of *Trypanosoma brucei* blood stream forms in a medium containing a low concentration of serum protein without feeder cell layers, *Journal of Parasitology*, **75**, 985-989

Hofheinz W., Merkli B. (1984) Quinine and Quinine Analogues, in: *Antimalarial drugs II*, Current Antimalarials and new drug development in W. Peters & W.H.G. Richard (eds), Springer Verlag, Berlin, Heidelberg, New York, Tokyo, p.61-81

Hudson A.T., (1984) Lapinone, menoctone, hydroxyquinolinquinones and similar structures, in W. Peters & W.H.G. Richard (eds), *Antimalarial drugs II*, Current Antimalarials and new drug development, Springer Verlag, Berlin, Heidelberg, New York, Tokyo, p.343-361

Hudson, A.T. (1993) Atovaquone-a novel broad spectrum antiinfective drug, *Parasitology today*, **9**, 60-66

Igweh, A.C., Onabanjo, A.O. (1989) Chemotherapeutic effects of *Annona senegalensis* in *Trypanosoma brucei brucei*, *Annals of Tropical Medicine and Parasitology*, **83**, 527-534

Iwu M.M., Jackson J.E., Schuster B.G. (1994) Medicinal plants in the fight against Leishmaniasis, Parasitlogy Today, 10, 65-68

Jackson, J.E., Tally, J.D., Tang, D.B. (1989) An in vitro micromethod for drug sensitivity testing of *Leishmania*, *American Journal of Tropical Medicine and Hygiene*, **41**, 318-330

Janse, C.J., Waters, A.P. (1995) *Plasmodium berghei*, The application of cultivation and purification techniques to molecular studies of malaria parasites, *Parasitology Today*, 11, 139-143

Jennings, F.W., Gray, G.D. (1983) Relapsed parasitaemia following chemotherapy of chronic *T. brucei* infections in mice and its relation to cerebral trypanosomes, *Microbiology and Immunology*, **7**, 147-154

Jensen, J.B., Trager, W. (1977) Plasmodium falciparum in culture, use of outdated erythrocytes and description of the candle jar method, *Journal of Parasitology*, **63**, 883-886

Kaminsky, R., Zwygarth E. (1989) Feeder Layer-free *in vitro* assay for screening antitrypanosomal compounds against *Trypanosoma brucei brucei* and *T.b. evansi*, *Antimicrobial Agents and Chemotherapy*, **66**, 881-885

Kelly, J.M., Das, P., Tomas, A.M. (1994) An approach to functional complementation by introduction of large DNA fragments into *Trypanosoma cruzi* and *Leishmania donovani* using a cosmid shuttle vector, *Molecular Biochemistry and Parasitology*, **65**, 51-62

Kirby G.C. (1996) Medicinal plants and the control of parasites, *Transactions of the Royal Society of Tropical Medicine and Hygiene*, **90**, 605-609

Li, J., Zhu, J.D., Appiah, A., McCutchan, T.F., Long, G.W., Milhous, W.K., Hollingdale, M.R. (1991) *Plasmodium berghei*, quantitation of *in vitro* effects of antimalarial drugs on exoerythrocytic development by a ribosomal RNA probe, *Experimental Parasitology*, **72**, 450-458

Li, R., Kenyon, G.L., Cohen, F.E., Chen, X., Gong, B., Dominguez, J.N., Davidson, E., Kurzban, G., Miller, R.E., Nuzum, E.O. (1995) *In vitro* antimalarial activity of chalcones and their derivatives, *Journal of Medicinal Chemistry*, **38**, 5031-5037

Lima, HC., Bleyenberg J.A., Titus R.G. (1997) A simple method for quantifying *Leishmania* in tissues of infected animals, *Parasitology Today*, **77**, 80-85

Lopes, J.N., Cruz-F.S., Docampo,R., Vasconcellos, M.E., Sampaio M.C., Pinto, A.V., Gilbert, B. (1978) *In vitro* and *in vivo* evaluation of the toxicity of 1,4-naphthoquinone and 1,2-naphthoquinone derivatives against *Trypanosoma cruzi*, *Annals of Tropical Medicine and Parasitology*, **72**, 523-531

Marshall R.J., Flanigan, T. (1992) Paromomycin inhibits *Cryptosporidium* infection of a human enterocyte cell line, *Journal of Infectious Diseases*, **165**, 772-774

McDonald, V., Stables, R., Warhurst, D.C., Barer, M.R., Blewett, D.A., Chapman, H.D., Connolly, G.M., Chiodini, P.L., McAdam, K.P. (1990) *In vitro* cultivation of *Cryptosporidium parvum* and screening for anticryptosporidial drugs, *Antimicrobial Agents and Chemotherapy*, **34**, 1498-1500

McFadden, D.C., Seeber, F., Boothroyd, J. C. (1997) Use of *Toxoplasma gondii* expressing b-galactosidase for colorimetric assement of drug activity *in vitro*, *Antibicrobial Agents and Chemotherapy*, **41**, 1849-1853

McHugh, T.D., Holliman R.E., Butcher P.D.(1994) The *in vitro* model of tissue cyst formation in *Toxoplasma gondii*, *Parasitology Today*, **10**, 281-285

McLaughlin, G.L., Deloron, P., Huong, A.Y., Sezibera, C., Campbell, G.H. (1988) DNA hybridization for assessment of response of *Plasmodium falciparum* to chloroquine therapy, *Journal of Clinical Microbiology*, **26**, 1704-1707

Mead J.A., You, X., Pharr, J.E., Belenkaya, Y., Arrowood, M.J., Fallon, M.T., Schinazi, R.F. (1995) Evaluation of Maduramicin and Alborixin in a SCID mouse model of chronic *Cryptosporidiosis*, *Antimicrobial Agents and Chemotherapy*, **39**, 854-858

Meshnick, S.R., Blobstein, S.H., Grady, R.W., Cerami, A. (1978) An approach to the development of new drugs for African trypanosomiasis, *Journal of Experimental Medicine*, **148**, 569-579

Meshnick, S.R., Taylor, T.E., Kamchonwongpaisan, S. (1996) Artemisinin and the antimalarial endoperoxides, from herbal remedy to targeted chemotherapy, *Microbiological Reviews*, **60**, 301-315

Navin, T.R., Arana, B.A., Arana, F.E., Berman, J.D., Chajon, J.F. (1992) Placebo-controlled clinical trial of sodium stibogluconate (Pentostam) versus ketoconazole for treating cutaneous leishmaniasis in Guatemala, *Journal of Infectious Diseases*, **165**, 528-534

Neal, R.A., Allen, S., McCoy, N., Olliaro, P., Croft, S.L. (1995) The sensitivity of *Leishmania* species to aminosidine, *Journal Antimicrobial Chemotherapy*, **35**, 577-584

Neal, R.A., Croft, S.L. (1984), An *in vitro* system for determining the activity of compounds against the intracellular amastigote form of *Leishmania donovani*, *Journal of Antimicrobial Chemotherapy*, **14**, 463-475

Nuzum, E., White-F., Thakur, C., Dietze, R., Wages, J., Grogl, M., Berman, J. (1995) Diagnosis of symptomatic visceral leishmaniasis by use of the polymerase chain reaction on patient blood, *Journal of Infectious Diseases*, **171**, 751-754

O'Neill, M.J., Bray, D.H., Boardman, P., Phillipson, J.D., Warhurst, D.C. (1985) Plants as sources of antimalarial drugs. Part 1. *In vitro* test method for the evaluation of crude extracts from plants, *Planta Medica* (5), 394-398

Obexer, W., Schmid, C., Brun, R. (1995) A novel *in vitro* screening assay for trypanocidal activity using the fluorescent dye BCECF-AM, *Tropical Medicine and Parasitology*, **46**, 45-48

Ofulla, A.V., Okoye, V.C., Khan, B., Githure, J.I., Roberts, C.R., Johnson, A.J., Martin, S.K. (1993) Cultivation of *Plasmodium falciparum* parasites in a serum-free medium. *American Journal of Tropical Medicine and Hygiene*, **49**, 335-340

Olliaro, P.L., Bryceson, A.D.M. (1993) Practical progress and new drugs for changing patterns of leishmaniasis, *Parasitology Today*, **9**, 323-328

Olliaro, P.L., Trigg, P.I. (1995) Status of antimalarial drugs under development, *Bulletin of the World Health Organisation*, **73**, 565-571

Peters, W. (1987), Chemotherapy and drug resistance in Malaria, *Vol. 2*, (2nd ed), Academic Press, London

Pfefferkorn, E.R., Pfefferkorn, L.C. (1977) Specific labeling of intracellular *Toxoplasma gondii* with uracil, *Journal of Protozoology*, **24**, 449-453

Philippson J.D., Wright C.W., Kirby G.C., Warhust D.C. (1995) Phytochemistry of some plants used in traditional medicine for the treatment of protozoal diseases, in K. Hostettmann, A. Marston, M. Maillard, M. Hamburger (eds), *Phytochemistry of plants used in traditional medicine*, Clarendon press, Oxford, pp: 96-135

Phillipson J.D., Wright C.W. (1991) Medicinal plants against protozoal diseases, *Transactions of the Royal Society of Tropical Medicine and Hygiene*, **85**, 18-21

Radloff, P.D., Philipps, J., Nkeyi, M., Sturchler, D., Mittelholzer, M.L., Kremsner, P.G. (1996) Arteflene compared with mefloquine for treating *Plasmodium falciparum* malaria in children, *American Journal of Tropical Medicine and Hygiene*, **55**, 259-262

Rochelle, P.A., Ferguson, D.M., Handojo, T.J., De Leon, R., Stewart, M.H., Wolfe, R.L. (1997) An assay combining cell culture with reverse transcriptase PCR to detect and determine the infectivity of waterborne *Cryptosporidium parvum*, *Applied Environmental Microbiology*, **63**, 2029-2037

Rogier, C., Brau, R., Tall, A., Cisse, B., Trape, J.F. (1996) Reducing the oral quinine, quinidine, cinchonin (Quinimax) treatment of uncomplicated malaria to three days does not increase the recurrence of attacks among children living in a highly endemic area of Senegal, *Transactions of the Royal Society of Tropical Medicine and Hygiene* **90**, 175-178

Sepulveda-Boza, S., Cassels, B.K. (1996) Plant metabolites active against *Trypanosoma cruzi, Planta Medica,* **62**, 98-105

Sinden, R.E., Suhrbier, A., Davies, C.S., Fleck, S.L., Hodivala, K., Nicholas, J.C. (1990) The development and routine application of high-density exoerythrocytic-stage cultures of *Plasmodium berghei, Bulletin of the World Health Organisation.* **68**, Suppl. l, 115-125

Sweeney T.R. (1984) Drugs with quinine-like action, in W. Peters & W.H.G. Richard (eds), *Antimalarial drugs II,* Current Antimalarials and New Drug development, Springer Verlag, Berlin, Heidelberg, New York, Tokyo

Tan-Ariya, P., Brockelman, C.R., Menabandhu,C. (1987) Optimal concentration of p-aminobenzoic acid and folic acid in the *in vitro* assay of antifolates against *Plasmodium falciparum, American Journal of Tropical and Medical Hygiene,* **37**, 42-48

Ter-Kuile, F., White, N.J., Holloway, P., Pasvol, G., Krishna, S. (1993) *Plasmodium falciparum- in vitro* studies of the pharmacodynamic properties of drugs used for the treatment of severe malaria, *Experimental Parasitology,* **76**, 85-95

Torres, R.A., Weinberg, W., Stansell, J., Leoung, G., Kovacs, J., Rogers, M., Scott, J. (1997) Atovaquone for salvage treatment and suppression of toxoplasmic encephalitis in patients with AIDS, *Clinical and Infectious Diseases,* **24**, 422-429

Urbina, J.A., Marchan, E., Lazardi, K., Visbal, G., Apitz-Castro, R., Gil, F., Aguirre, T., Piras, M.M., Piras, R. (1993) Inhibition of phosphatidylcholine biosynthesis and cell proliferation in *Trypanosoma cruzi* by ajoene, an antiplatelet compound isolated from garlic, *Biochemistry and Pharmacology,* **45**, 2381-2387

Van Vianen, P.H., Thaithong, S., Reinders, P.P., Van Engen, A., Van der Keur, M., Tanke, H.J., Van der Kaay, H.J., Mons, B. (1990) Automated flow cytometric analysis of drug susceptibility of malaria parasites, *American Journal of Tropical Medicine and Hygiene,* **43**, 602-607

Vennerstrom, J.L., Lovelace, J.K., Waits, V.B., Hanson,W.L., Klayman, D.L. (1990) Berberine derivatives as antileishmanial drugs, *Antimicrobial Agents and Chemotherapy,* **34**, 918-921

Verdine, G.L. (1996) The combinatorial chemistry of nature, *Nature,* **384**, Suppl 1, 11-13

Vial, H. (1996) Recent developments and rationale towards new strategies for malarial chemotherapy, *Parasite,* **3**, 3-23

Warhurst D.C. (1997) Drug resistants in Malaria-Laboratory and field investigations, *Journal of Pharmacy and Pharmacology,* **49**, Suppl. 2, 3-7

Weiss, C.R., Brun R., Freiburghaus F., Schaffner,W. (1995) Antitrypanocidal potency of medical plants used in traditional medicine in Ivory Coast, *43rd Annual Congress on Medicinal Plant Research,* Posterpresentation 56

Wernsdorfer, W.H. (1994) Epidemiology of drug resistance in malaria, *Acta Tropica,* **56**, 143-156

White, N.J. (1994) Artemisinin-current status, *Transactions of the Royal Society of Tropical Medicine and Hygiene* **88**, Suppl. 1, 3-4

WHO (1984) Advances in Malaria chemotherapy, *Technical Report Series,* World Health Organisation, Geneva, p. 35

Wright, C.W., Phillipson J.D. (1990) Natural products and the development of selective antiprotozoal drugs, *Phytotherapy Research,* **4**, 127-129

Zinsstag, J., Brun, R., Gessler, M. (1991) A new photometric assay for testing trypanocidal activity *in vitro, Parasitology Research,* **77**, 33-38.

IN-VITRO ASSAYS FOR ACTIVITY-GUIDED ENRICHMENT OF IMMUNOMODULATORY PLANT CONSTITUENTS

H. VAN DIJK*, C.J. BEUKELMAN, B.H. KROES, S.B.A. HALKES, H.F. SMIT, LINDA C. QUARLES VAN UFFORD, E. VAN DEN WORM, TINEKE L. TINBERGEN-DE BOER, J.H. VAN MEER, A.J.J. VAN DEN BERG & R.P. LABADIE
Utrecht University Department of Pharmacy, Section of Medicinal Chemistry, Pharmacognosy Unit, F.A.F.C. Went Building, Sorbonnelaan 16, NL-3584 CA Utrecht, The Netherlands

It has been known for ages or even millennia that certain plants or plant preparations may be used successfully to selectively treat immunological disorders. In the course of the last centuries, it has become clear that the active principle(s) of such 'immunomodulatory' plants/plant preparations may be single chemical entities or more complex mixtures of related substances that can either enhance or suppress deranged immunological reactions. Depending on their mode of action, plant-derived 'immunomodulators' can be used to stimulate the immune system of immuno-compromised individuals (patients with congenital or acquired immunodeficiencies, young children, or elderly people) or, alternatively, to suppress the immune system of hyperreactive subjects (patients with allergic, autoimmune, and/or rheumatic diseases) or transplantation patients. There are even examples of plant-derived substances with more or less selective anti-lymphoproliferative effects.

During the last fifteen years, our approach towards recognizing the active principles of immunomodulatory plants or plant preparations has been the activity-guided biochemical isolation and subsequent identification or characterization of the active constituents. This approach is based on Prof. Labadie's 'biomarker' concept that the presence of biologically active constituents in plant extracts/preparations could be tested first using simple, rather nonselective *in-vitro* assays and could be assessed later with more selective or even specific assays.

In our attempts to search for immunomodulatory plant constituents, the assays we used to select active fractions have been based on *in-vitro* parameters of innate immunity, specific immune responses, and systemic or mucosal immunological adjuvant activity. The background of these techniques is discussed and some results are presented.

*To whom correspondence should be addressed.

1. Introduction

The immune system is a complex, highly coordinated system involved in defending the body not only against microbial invaders, but also against malignant cells generated by the body itself. Although the system is very well regulated, different derangements may occur resulting in, for example, autoimmunity, rheumatic diseases, or allergic reactions. Even leukemias and lymphomas may be considered as dysregulations of the immune system. In addition to these 'inflicted' derangements, hereditary defects in the immune system are known.

Some of the beneficial effects of traditionally used phytomedicines can be explained by their effect on the immune system. In order to facilitate the standardization of immunologically active phytotherapeutics, our research group has developed a strategy of activity-guided isolation, identification, and/or characterization of immunologically active constituents, which together form the basis of biological and sometimes even physico-chemical standardization of plant preparations. This strategy is in fact the practical elaboration of the 'biomarker' concept (Labadie, 1990; Van Dijk et al., 1994).

TABLE 1. The biomarker concept

- Basis: ethnopharmacological use
- Use of simple, relevant technique:
  are bioactive constituents present?
- Activity-guided isolation
- Identification/characterization of active constituents
- Development of specific assay for standardization purposes

The biomarker concept is based on the assumption that the phytomedicine under study is biologically active (step one). If this assumption is correct, the investigator should look for one or more simple, but relevant *in-vitro* test systems which can be used to roughly estimate the biological activity (step two). The third step is the activity-guided, biochemical isolation of the active constituent(s), after which the active principle(s) can be identified or at least characterized (e.g. with polysaccharides) (step four). The final step and ultimate goal of the biomarker concept is the development of (a) more selective or even specific test system(s), which can be applied to standardize the phytomedicine in question or related preparations (step five).

2. The immune system (first section)

As stated above, the immune system is aimed at defending the body against foreign invaders. There are various connections between the immune system and other physiological systems, the pituitary-adrenal gland axis, the reproductive system, etc. It has even been suggested that the immune system may dictate the time of delivery. The immune system is considered to act in different 'lines of defense', starting with defense at the epithelial barrier (Table 2).

TABLE 2. The immune system

- ***Defense at the epithelial barrier***
- ***Nonspecific defense mechanisms***
  - humoral defense mechanisms
    → cytokine network
    → complement system
  - cellular compartment
    → polymorphonuclear cells (PMN)
    → mononuclear phagocytes (Mϕ)
    → natural killer (NK) cells
- ***Antigen-specific immunological reactions***
  - T-cell activation and cytokine profiles
  - B-cell activation and antibody formation

The epithelium is colonized by many different microorganisms which adhere to surface-exposed cells and from which site they may start an infection. In our late antibiotic era, it can be envisaged that the interaction between microorganisms and their acceptor sites on epithelial cells could become an increasingly important target for novel antimicrobial drugs, also those from natural resources.

The second line of defense comprises the nonspecific humoral defense mechanisms, cytokines and complement with all their effects (the induction of cell differentiation, the mobilization and priming/activation of inflammatory cells, and the killing of Gram-negative bacteria and virusses). The cellular compartment of the second line of defense consists of phagocytic cells (polymorphonuclear cells; PMN), monocytes/macrophages (mononuclear phagocytes; Mf), and the nonphagocytic 'natural killer' (NK) cells. As far as the *cytokine network* is concerned, the Mf-derived cytokines interleukin-1 (IL-1), IL-6, and tumor-necrosis factor alpha (TNFa) together are responsible for the most important events in the septic shock syndrome induced by either Gram-negative or Gram-positive microorganisms. The similarity between the activation pathways in mononuclear phagocytic cells triggered by Gram-positive and Gram-negative bacteria (Mattsson et al., 1996) makes these pathways useful targets or 'receptors' for the development of novel drugs. Other cytokines and cytokine receptors (e.g. IL-12 and IL-4R) are interesting subjects for generating novel agonists or blocking agents that can be used to modulate deranged cytokine-regulated immunological reactions (rheumatoid arthritis, asthma, etc.).

The *complement system* includes complex cascades of activating and regulatory principles involved in the early recognition of non-self and the initiation of the inflammatory reaction. One of the major activation pathways remains the so-called 'classical' complement pathway, which is dependent on antibodies and results in the rapid generation of chemotactic factors (C5a and Ba), the mobilization of PMN to the primary site of activation (infection), the coating of foreign bodies with C3b molecules which stimulate phagocytosis (opsonization), and, finally, the enhanced uptake and killing of microorganisms by PMN and, later, by Mϕ. An important common feature of complement-inhibitory compounds is that they possess immunological adjuvant activity (Klerx, 1985), i.e. they may be used as an additive to antigen preparations in order to enhance the specific immune response against them (e.g. with vaccines). Since adjuvants are usually quite simple and inexpensive chemical compounds, they can be used to decrease the dose and thereby contribute to a reduced toxicity and a lower price of vaccines.

In the next section, we will describe the use of the classical complement pathway in human serum as a target to find anticomplementary components in extracts of the plant *Filipendula ulmaria* (L.) Maxim. (queen of the meadow; meadowsweet). We use this example to illustrate different aspects of the biomarker concept.

3. Anticomplementary activity of Filipendula ulmaria extracts

F. ulmaria (meadowsweet) is a perennial herb, endemic in Europe and Asia, but also found in Northern America. In traditional medicine, this herb has been used in the treatment of inflammatory diseases including gout and rheumatoid arthritis (Madaus, 1938; Gessner and Orzechowski, 1974). Different flavonoids, terpenoids, and salicylic-acid derivatives have been isolated from different parts of the plant (Meier et al., 1987; Scheer and Wichtl, 1987; Valle et al., 1988; Lamaison et al., 1992).

To substantiate the claimed anti-inflammatory effects (first step in the biomarker concept), different extracts of the roots, herbs, and flowers of *F. ulmaria* were prepared and tested in assays for anticomplementary and anti-inflammatory activity (Halkes et al., 1997a) (step two). The extracts were obtained by Soxhlet extraction using an eluotropic series of organic solvents, ranging from strongly apolar (light petroleum) to hydrophilic (water).

In this example, the anticomplementary effects of *F. ulmaria* flower extracts in the classical complement activation pathway were studied (Halkes et al., 1997b). As shown in Table 3, all *F. ulmaria* flower fractions inhibited the human classical complement pathway, with the ethyl acetate fraction being the most potent one. The methanol fraction showed a much higher yield, but a 4-5-times lower activity. Since the purpose of this study was not to optimize the purification scheme, but just to achieve the identification or characterization of active *F. ulmaria* constituents, the ethyl acetate fraction was subjected to further fractionation (step three). A Sephadex LH20 column was used for this purpose. Eighty different fractions were obtained using this column, each of which was separately tested for inhibitory activity in the assay for human classical complement pathway activation.

Fractions 10-24 and 42-48 showed significant antihemolytic activity, which could not be reproduced in an assay for human alternative complement pathway activation (not shown). This implies that the active fractions interfere with the classical complement pathway. Of these fractions, 10-24 showed by far the highest specific inhibitory activity (Fig. 1). Although they were not fully characterized (step four) yet, the active fractions were

TABLE 3. Yields and specific activities of serial Soxhlet extracts from *F. ulmaria* flowers on the classical activation pathway of human complement

Solvents	Yield (%)	Specific activity (units/mg)
Light petroleum	1.6	555 ± 37
Diethyl ether	0.7	680 ± 42
Ethyl acetate	2.9	2300 ± 238
Methanol	33.6	520 ± 33
Water	8.9	130 ± 12

pooled and the pool was further used to devise a more specific assay to estimate the anticomplementary activity of *F. ulmaria* extracts. To this end, the classical complement pathway assay was adapted in such a way that four different assays were developed. One single classical complement pathway component was the limiting factor in each assay. These assays showed that complement component C1, but not C2, C3, or C4, was inhibited by the pooled *F. ulmaria* Sephadex LH20 fraction (Fig. 1), indicating that the C1 assay is probably the assay of choice for the standardization of the anticomplementary activity of ethyl acetate *F. ulmaria* Soxhlet extract according to the biomarker principle (Table 1) (step five).

Fig.1. Anticomplementary effects of ethyl acetate *F. ulmaria* Soxhlet extract separated on a Sephadex LH20 column (Y (%) represents yield; U/mg, specific anticomplementary activity)

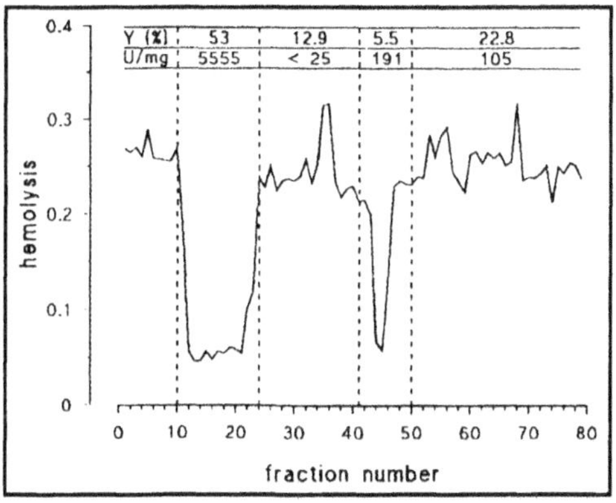

Fig. 2. Anticomplementary behavior of pooled Sephadex LH20 10-24 fractions of *F. ulmaria* ethyl acetate Soxhlet extract under conditions where complement components C1, C2, C3, or C4 were the limiting factors

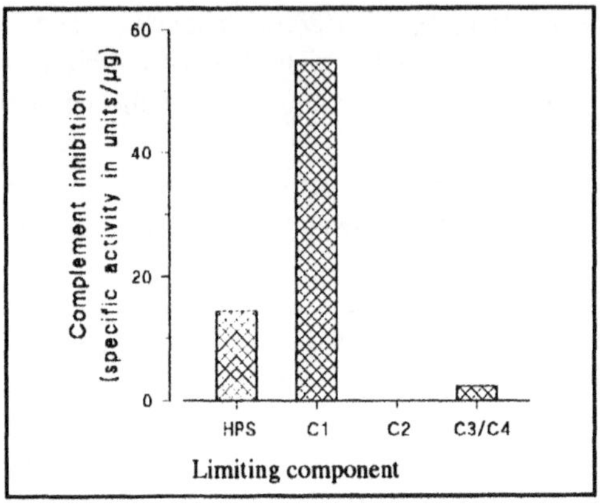

4. The immune system (continued)

As already mentioned, cells participating in the nonspecific cellular defense include phagocytes (PMN), mononuclear phagocytes (monocytes and Mϕ), and the nonphagocytic NK cells. These three cell types have all different granules (lysosomes) which play an important role in the phagocytosis and killing processes. Lysosomes are important targets when trying to interfere with the function of phagocytes and NK cells, since lysosomotropic agents can influence the pH of the granules and thereby the functions of the cells (Vandenbroucke-Grauls et al., 1984).

One important extralysosomal system within *PMN* is NADPH oxidase, which can be assembled from different membrane-bound and cytosolic components when the cells are primed/triggered (Fig. 3). NADPH oxidase uses oxygen as a substrate and gives rise to the generation of superoxide anion (O_2- radical) inside the granules and outside the cells. This short-lived intermediate decomposes to hydrogen peroxide, which can form hydroxyl radicals (OH◊) and, together with chloride anions, hypochlorite; this latter reaction is catalyzed by myeloperoxidase (MPO). The decay of these oxygen species is accompanied by the generation of photons which can be visualized using different enhancers such as luminol (whole spectrum) or lucigenin (nonhypochlorite species). When studying the anti-inflammatory Indian and Sri Lankan plant *Picrorhiza kurroa*, we found a very active component which could be identified as apocynin (4-hydroxy-3-methoxy-acetophenone; Simons et al., 1989). Apocynin has a very remarkable mode of action: it interferes with the assembly of the membrane-bound NADPH oxidase ('t Hart et al., 1990). Our results implicate that the oxygen-species-generating system of PMN is an important 'receptor' to find new lead compounds with anti-inflammatory potential.

Mononuclear phagocytes (monocytes and Mf) are supposed to generate oxygen species to a lower level than PMN, but they additionally have a completely different radical-generating system called nitric oxide synthase (NOS; Fig. 4). Two different NOS types are known, a constitutive (cNOS) and an inducible (iNOS) type (Nathan and Xie, 1994). Both

Fig. 3. The generation of oxygen radivals by primed PMN

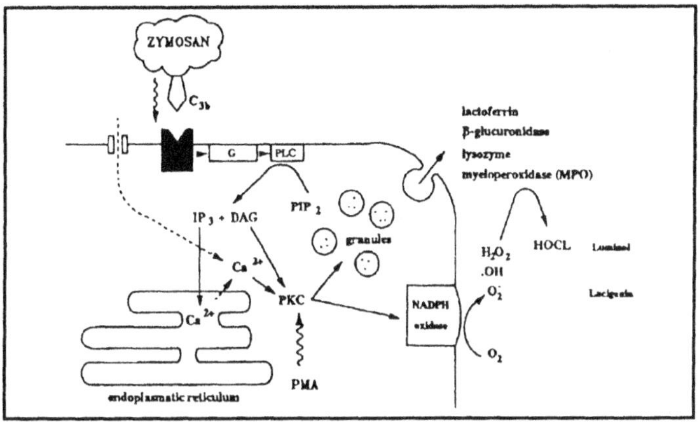

enzymes deaminate the amino acid L-arginine to citrulline and oxidize the amine to nitric oxide radical (NO◊). This nitric oxide is a very potent agent which is held responsible for the antimicrobial activity of mononuclear phagocytes (Moncada, Palmer, and Higgs, 1991; Schmidt and Walter, 1994). Substances that compete with arginine for binding site on NOS, e.g. N-methylated analogues, inhibit NO◊ generation. This makes the NO-generating systems important targets for the development of novel anti-inflammatory drugs, including plant-derived antiphlogistics. Recent publications suggest that the reaction product of hydrogen peroxide and NO◊, peroxynitrite, is even more toxic than NO◊ itself (Beckman and Koppenol, 1996; Muijsers et al., 1997).

Fig. 4. The nitric-oxide-generating systems of mononuclear phagocytes

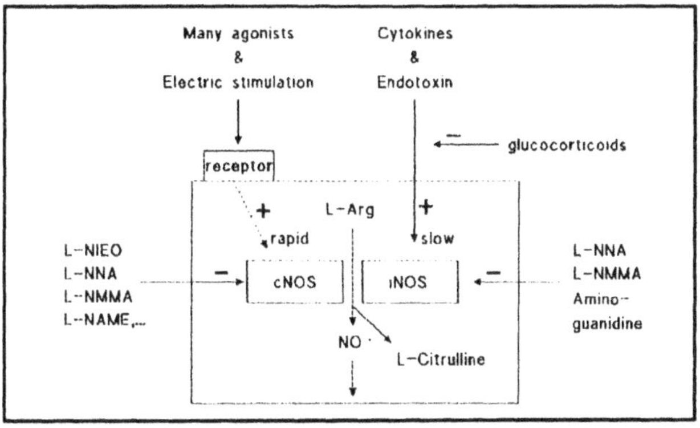

The role of NK cells in human disease is not yet fully understood. Textbooks claim that they may be involved in antitumor and probably also antiviral defense and recent papers

suggest their contribution to antiprotozoan defense (Scharton-Kersten and Sher, 1997). Although we do not exclude a role in, for example, autoimmune phenomena, the NK cell is not an important item in novel-drug development at present. With regard to antigen-specific immunological reactions, we would like to make a distinction between systemic ('conventional' or 'central') immunological reactions on the one hand, that are brought about by thymus-derived, αβ-positive, and CD4- or CD8-positive T cells and bone-marrow-derived B cells and, mucosal immune responses on the other, which include less 'conventional' T cells with absent or aberrant CD4 or CD8 molecules on their surface, a T-cell type that lacks the αβ-type antigen receptor and has a γδ-type antigen receptor instead, and B cells that, upon stimulation, may generate dimeric or trimeric-IgA-producing plasma cells. There are reasons to suppose that these 'mucosal' T and B cells are of other origin and have a specific function, i.e. the protection of epithelial surfaces.

5. The systemic immune system

Systemic (central, conventional) immunological reactions are directed against antigens that enter the body either via the skin or via invasive microorganisms. Where possible, mononuclear phagocytes in subcutaneous connective tissue will phagocytose the antigens, take them to the nearest draining lymph node, and present antigen-derived epitopes to CD4-positive T cells with the right antigen receptor, which will start a primary cellular reaction. Depending on the cytokine profiles produced locally, the responding T cells will continue to react and secrete cytokines which may stimulate other cells, i.e. other CD4-positive T cells, CD8-positive T cells, and/or B cells to form memory cells and, in the case of B cells, antibody-producing plasma cells. The first reacting T cells can also leave the draining lymph node and, via the lymph vessels arrive in the bloodstream again. From the blood, they may easily leave postcapillary venules and reenter tissues where they come in contact with antigen and/or antigen-presenting Mϕ. Now, the T cells will start secreting interforon gamma (IFNγ) which mobilizes new mononuclear phagocytes from the bloodstream and activates these phagocytes so that they can eliminate antigens taken up by the phagocytes. Upon systemic entry of the antigen (pathogenic microorganisms), similar processes may take place in the spleen. Primary B cell activation mostly results in the production of IgM antibodies, whereas memory B cells may synthesize IgG or IgE.

The activation of systemic T and B cells is a lymphoproliferative process that, in the case of antimicrobial immune responses, can be rather efficiently enhanced by immunological adjuvants or, in case of autoimmune or rheumatic diseases, be blocked by e.g. cytostatic agents. The way to discover such immunomodulatory agents is to use *in-vitro* models of T- and/or B-cell stimulation with limiting amounts of stimuli that cause either polyclonal activation of the cells (phytohemagglutinin; poke-weed mitogen) or restricted superantigenic stimulation. These systems can thus be used to find novel (co)stimulatory entities or immunosuppressive agents.

6. The mucosal immune system

As the term indicates, the mucosal immune system is associated with the mucosal surfaces that cover the gut, the respiratory tract, the urogenital tract, the gall bladder, the mammary glands, the vagina, and the conjunctivae. As far as we know now, the most important entry sites for antigenic stimulation of the mucosal immune system are specialized epithelial cells in the nose and small intestine, that in the gut are called M cells. The antigen may be transferred from these sites to local mucosal lymph nodes (e.g. Peyer's patches in the small intestine) and stimulate the T and/or B cells present there (Fig. 5). Although the exact roles of the different lymphocytes in the mucosal immune system are not yet fully understood, it

Fig. 5. A schematic view of inducing and efferent pathways in IgA production and excretion

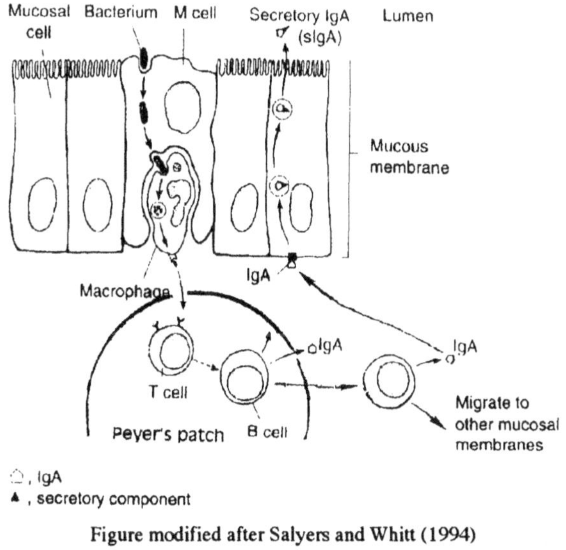

Figure modified after Salyers and Whitt (1994)

is known that IgA-producing B cells leave the local lymph node at a certain time after antigenic stimulation and travel via the lymph vessels and the bloodstream to distant mucosal surfaces where they continue to secrete oligomeric IgA molecules. Like IgM antibodies, these IgA molecules bear a J chain, which is recognized by a receptor on the basal side of epithelial cells. The epithelial cells subsequently pinocytose the IgA molecules and transport the resultant vesicles to the apical side of the cells, from where the IgA molecules are then exocytosed. The whole process of pinocytosis, transport of vesicles from the basal to the apical side, and the exocytosis of IgA molecules is referred to as the transcytosis of IgA. An enzyme is released at the apical side of the epithelial cells which is able to split off the larger part of the J-chain receptor and separates, in turn, the transcytosed IgA from the epithelial cell. This IgA then becomes 'secretory' IgA in the gut or tract. The part of the released J-chain receptor, which is also known as 'secretory component', now covers the Fc portions of the IgA subunits and protects the secretory IgA molecules from acid hydrolysis or alkaline digestion in the stomach and gut. This is why IgA from breast milk is also resistant to hydrolysis in the gut of sucklings and can protect the gut of young children against water-borne pathogens.

The critical step in the stimulation of IgA formation and excretion at mucosal surfaces seems to be the receptor in the nose and on M cells in the gut which is involved in the uptake of antigens by the mucosal immune system. Since cholera bacteria and enterotoxigene *Escherichia coli* bacilli have an intrinsic adjuvant for mucosal immune responses interacting with this receptor, the identity of this receptor has been easily established. It is a simple glycolipid in the epithelial-cell membrane called ganglioside Gm1 and ganglioside Gm2, for mice and man, respectively. This type of receptor may become an important tool in discovering new lead compounds for immunomodulation at mucosal surfaces (Kato and Owen, 1994; Russell-Jones, 1994).

7. Concluding remarks

In this contribution, we have discussed the complexity of the immune system and different existing or novel targets for the identification of lead compounds of plant origin with immunomodulatory activity. We hope, that our approach may stimulate colleagues in the field to join us in this ever-fascinating field of research.

8. References

Beckman, J.S. and Koppenol, W.H. (1996) Nitric oxide, superoxide, and peroxynitrite: the good, the bad, and the ugly. *Am. J. Physiol.* **271**, C1424-1437.

Gessner, O. and Orzechowski, G. (1974) *Gift- und Arzneipflanzen von Mitteleuropa,* 3rd ed., Carl Winter Universit≥tverlag, Heidelberg, 182-184.

Halkes, S.B.A., Beukelman, C.J., Kroes, B.H., Van den Berg, A.J.J., Labadie, R.P., and Van Dijk, H. (1997a) In vitro immunomodulatory activity of the roots, herb, and flowers of *Filipendula ulmaria. Phytother. Res.* 11: in press.

Halkes, S.B.A., Beukelman, C.J., Kroes, B.H., Van den Berg, A.J.J., Van Dijk, H., and Labadie, R.P. (1997b) A strong complement inhibitor from the flowers of *Filipendula ulmaria* (L.) Maxim. *Pharmaceut. Pharmacol. Let.*, in press.

Kato, T. and Owen, R.L. (1994) Structure and function of intestinal mucosal epithelium. In: *Handbook of Mucosal Immunology* (P.L. Orga, J. Mestecky, M.E. Lamm, W. Strober, J.R. McGee, and J. Bienenstock, eds.). Academic Press, San Diego, USA, pp. 11-26.

Klerx, J.P.A.M. (1985) Immunological adjuvant activity: complement-dependent and complement-independent processes. *Thesis*, Utrecht, The Netherlands.

Labadie, R.P. (1990) Exploring traditional medicine - A challenge for innovations in drug development. Proceedings of the *International Symposium on Innovations in Pharmaceutical Sciences and Technology* (Oct. 27-29, 1990; Sarkhej, India) pp. 122-131.

Lamaison, J.L., Petitjean-Freytet, C., and Carnat, A. (1992) Teneur en principaux flavonoïdes des parties aériennes de *Filipendula ulmaria* (L.) Maxim. subsp. *ulmaria* et subsp. *denudata* (J. & C. Presl) Hayek. *Pharm. Acta Helv.* **67**, 218-222.

Madaus, G. (1938) *Lehrbuch der biologischen Heilmittel*, Vol. 3, Georg Thieme Verlag, Leipzig, 2593-2597.

Mattsson, E.E., Van Dijk, H. Van Kessel, K., Verhoef, J., Fleer, A., and Rollof, J. (1996) Intracellular pathways involved in tumor necrosis factor-α release by human monocytes upon stimulation with LPS or staphylococcal peptidoglycan are partly similar. *J. Infect. Dis.* **173**, 212-218.

Meier, B., Lehmann, D., Sticher, O., and Bettschart, A. (1987) Salicylate in Arzneipflanzen. *Dtsch. Apoth. Ztg.* **46**, 2401-2407.

Muijsers, R.B.R., Folkerts, G., Henricks, P.A.J., Sadeghi-Hashjin, G., and Nijkamp, F.P. (1997) Peroxynitrite: a two-faced metabolite of nitric oxide (minireview). *Life Sci.* **60**, 1833-1845.

Moncada, S., Palmer, R.M.J., and Higgs, E.A. (1991) Nitric oxide: physiology, pathophysiology, and pharmacology. *Pharmacol. Revs* **43**, 109-142.

Nathan, C. and Xie, Q.-W. (1994) Nitric oxide synthases: roles, tolls, and controls. *Cell* **78**, 915-918.

Russell-Jones, G.J. (1994) Oral vaccination with lectins and lectin-like molecules. In: *Novel Delivery Systems for Oral Vaccines* (D.T. O'Hagan, ed.), CRC Press, Ann Arbor, pp. 225-240.

Salyers, A.A. and Whitt, D.D. (1994) *Bacterial Pathogenesis; a Molecular Approach. ASM Press*, Washington DC, USA, p. 9.

Scharton-Kersten, T.M. and Sher, A. (1997) Role of natural killer in innate resistance to protozoan infections. *Curr. Opin. Immunol.* **9**, 44-51.

Scheer, T. and Wichtl, M. (1987) Zum Vorkommen von Kämpferol-4'-O-β-D-glucopyranosid in *Filipendula ulmaria* and *Allium cepa*. *Planta Med.* **53**, 573-574.

Schmidt, H.H.H.W. and Walter, U. (1994) NO at worl. *Cell* **78**, 919-925.

Simons, J.M., 't Hart, L.A., Labadie, R.P., Van Dijk, H., and De Silva, K.T.D. (1990) Modulation of human complement activation and the human neutrophil oxidative burst by different root extracts of *Picrorhiza kurroa*. *Phytother. Res.* **4**, 207-211.

't Hart, B.A., Simons, J.M., Rijkers, G.T., Hoogvliet, J.C., Van Dijk, H. & Labadie, R.P. (1990) Reaction products of 1-naphtol with reactive oxygen species prevent NADPH oxidase activation in activated human neutrophils, but leave phagocytosis intact. *Free Radical Biol. Med.* **8**, 241-249.

Valle, M.G., Nano, G.M., and Tira, S. (1988) The essential oil of *Filipendula ulmaria*. *Planta Med.* **54**, 181-182.

Vandenbroucke-Grauls, C.M.J.E., Thijssen, H.M.W.M., Marcelis, J.M., Sharma, S.D., and Verhoef, J. (1984) Effects of lysosomotropic amines on human polymorphonuclear leukocyte function. *Immunology* **51**, 319-326.

Van Dijk, H., Beukelman, C.J., Quarles van Ufford, L.C., Kroes, B.H., Van den Berg, A.J.J., and Labadie, R.P. (1994) Phytomedicines and the immune system. *Eur. Phytotelegram* **6**, 14-17.

SEARCH FOR POTENT IMMUNOMODULATORY AGENTS FROM PLANTS AND OTHER SOURCES

H. WAGNER
Institute of Pharmaceutical Biology, University of Munich
Karlstrasse 29, D-80333 Muenchen, Germany

1. Introduction

Immunostimulants are leading predominantly to a nonspecific stimulation of immunological defence mechanisms. Nonspecific stimulants do not affect immunological memory cells and since their pharmacological efficacy fades comparatively quickly, they have to be administered either at intervals or continuously [1,2]. Different are the *immunoadjuvants, i.e.* complete or incomplete Freund's adjuvant, which are added to antigenes (vaccines) and hereby increase the production of antibodies without acting as antigenes themselves [3].

Why do we need immunostimulants? The major concern for developing and applying immunostimulants must be directed to restore the chronically suppressed immune system. The immunosuppression can be caused by bacterial and viral infections, cancer, environmental agents (*i.e.* pesticides or allergenes), excessive chemo- or radiotherapy, malnutrition or psychic stress. There is a concurrent under the experts tha an effectively working immunesystem is a requirement for full protection against infect nd the various sequels. In particular, immunostimulants are an attractive alternative entional chemotherapy when mixed infections, infectious hospitalism, chronic infectious diseases, persistent infections and resistent bacterial or viral infections have to be treated.

2. Screening methods

The result of more than 15 years of screening experience in the author's laboratory has shown that the best short-cut to find effective immunostimulants in plants are investigations using experimental animal models. The infectious stress or immunsuppression models have first preference. They indicate whether or not, and to what extent a drug is able to antagonize a severe, or otherwise lethal infection, and hereby show the protective or therapeutic potential of a drug. The immunosuppression model reveals to what extent a drug under test is able to restore an impaired or unbalanced immune system.

For mass screening, however, the various *in vitro* bioassays are the adequate methods, since at the same time they provide hints on the possible mechanism of action of a drug [4]. Since the results obtained *in vitro* not necessarily must have a counterpart *in vivo*, it is necessary to confirm them by *in vivo* experiments. *Vice versa* it is possible that positive *in vivo* results cannot be corroborated in an *in vitro* assay, when several cooperating cells or mediator systems nonpresent in an *in vitro* cell system are responsible for the *in vivo* effect.

In general, since there is no master or key cell in the immune system which governs and regulates all various immune reactions it is often necessary to carry out several *in vitro* assays.

2.1 *IN VIVO* ASSAYS

2.1.1 *Infectious stress assays with mice* [5,6]

These assays evaluate whether, and to what extent, a claimed immunostimulating agent is able to protect mice against an otherwise lethal infection.

The drug to be tested is administered *i.v.* to mice (day -1), 24 hrs (day 0) the animals receive an *i.v.* injection of a pathogenic microorganism (*i.e. Candida albicans, Listeria* monocytogenes). A second injection of the immunostimulant is administered on day +1. The protective effect can be measured by either

- a follow-up registration of the survival or death rate after a defined time,
- a continuous estimation of blood parameters (counts of leukocytes, T- and B-lymphocytes etc.) of pretreated and infected mice,
- removal of spleen, liver or kidney from mice, homogenization and transfer of an aliquot amount to agar plates and counting the number of CFU (colony forming units).
  Note: Listeria monocytogenes infections are mainly macrophage dependent whereas the infection load with *Candida* is primarily controlled by granulocytes.
  The above described investigations can be carried out with mice immunocompromized by treatment with cyclosporin A, cyclophosphamide or radiation. This treatment suppresses the function of the T-lymphocyte system.

2.1.2 *In vivo phagocytosis assay* [7]

This test is carried out with mice. The substance under test is administered *i.p.* or orally to mice. After 24 hrs each mouse receives an *i.v.* injection of 0.3 ml of Indian ink dispersion (colloidal carbon particle/30g bodyweight). Blood samples are taken from the retro-orbital venous plexus at intervals of 3, 6, 9, 12 and 15 min. after *i.v.* injection. The carbon clearance, *i.e.* the rate of elimination of the carbon from the blood, is determined by turbidimetric spectrophotometry at 650 nm. Density reading, plotted against time on a logarithmic scale, gives the regression lines. The stimulation rate is obtained as the ratio of the mean regression coefficient of the substance (R_{Ctr}) to the regression coeffcient of the control (R_{Cc}).

3. *In vitro* assays [4]

Among the variety of *in vitro* tests available today are those that allow the determination of the functional state and efficiency of the mononuclear phagocyte system. Ranked second are tests that measure the influence of compounds on T-lymphocyte populations and the complement system.

3.1 GRANULOCYTE SMEAR TEST [8]

The former granulocyte smear test of Brandt [8] has been replaced by the flow cytometric assay using whole blood samples and latex particles [9]. The phagocytosis index is calculated from the portion of cells that underwent phagocytosis as compared to that of the control. This test can be also carried out with peritoneal or bone marrow macrophages.

3.2 THE CHEMILUMINESCENCE ASSAY [10]

It measures the quantity of oxygen radicals produced during granulocyte or macrophage phagocytosis using luminol or lucigenol for measuring photons with a photomultiplier. Zymosan, PAF, TPA or calciumionophore can be used as a challenge.

3.3 CYTOKINE INDUCTION ASSAYS [11,12]

Cytokines (e.g. Il-1, Il-2, IFN-β_2-TNF-α) released from various immunocompetent cells after activation play a great role in the regulation and potentiation of the nonspecific and specific immune system. The cytokine induction tests can be performed directly with macrophages or thymocytes (measuring the concentration of released cytokines by specific antibodies) or indirectly (using the ^{3}H-tymidine incorporation method).

3.4 IMMUNE INDUCED CYTOTOXICITY ASSAY [5,11]

Macrophages can be transformed into effector cells by induction and stimulation to release cytotoxic effector substances. One of the most important factors is the membrane-bound tumor necrosis factor (TNFα) which necrotizes or lyses tumor cells. There are several modifications of these tests. In principle they vary according to whether the tumor cells are radio labeled with ^{51}chromium or ^{3}H-tymidine.

3.5 LYMPHOCYTE ASSAYS

T- and B-lymphocytes belong to the specific cellular immune system. They play an important role in practically all defence mechanisms. They can be activated directly by antigens and mitogens or indirectly by cytokines released from macrophages and granulocytes.

The most used assays are the T-lymphocyte proliferation assay [4] and the CD69 antigen expression-test [13].

3.6 COMPLEMENT ASSAYS [13,14]

The complement system, which belongs to the nonspecific humoral defence system, is involved in the defence against viruses and tumors. It can be subdivided into the classical and alternative activation pathway. The classical pathway is dependent on antibodies (IgM and IgG) whereas the alternative pathway can be activated by microorganisms on their own or in combination with IgA antibodies. Interference of drugs with the complement system results in the lysis of erythrocytes and the release of hemoglobin. The measurement of the hemolytic complement activity can be used for the assay in the classical (CP) as well as in the alternative (AP) complement pathway.

4. Selection of plants and isolated compounds for screening [2]

Since the term immunostimulation is new and emerged only recently, other indications described in the ethnomedicinal literature have to be used. According to more than 15 years, screening experience, plant drugs recommended for the treatment of bacterial, viral and parasidal infections, cancer, airway and skin diseases might be suitable candidates for screening. The following example may illustrate this strategy. The bark of *Tabebuia avellanedae* (Lapacho), used in folksmedicine for the treatment of cancer, contains naphthoquinones having a marked *in vivo* antitumoral effect. The same compounds, however, act *in vitro* as immunostimulants when applied in minute concentration (ng or pg range). This dose-dependent reversal effect of known cytotoxic compounds has been observed in our screeing program for more than 50% of the compounds tested [16]. These findings are consistent with the observation that substances which earlier were used for the so called ,shock therapy´ and are characterized as irritants or inflammatory agents such as phorbolesters (croton oil), pungent agents, mineral oils, saponins or sesquiterpenlactones, exert a marked immunostimulatory activity in low doses. In our screeing program some bitter substances, diacylglyzerol derivatives and protein- or tyrosinkinase-C-inhibitors also have been found to be potent immunostimulants.

5. Screening results

The compounds found worthy of more in-depth immunological investigations can be subdivided into the classes of low and high mol. weight compounds (see Table). They have already been listed in review articles [1,2,17]. In the class of low molecular weight compounds the bryostatins from the marine organism *Bugula neritina* [18] and naphthoquinones [16] may be of particular interest, whereas in the class of high molecular compounds some polysaccharides and lectins from higher plants and microorganisms might be promising candidates for clinical studies.

TABLE. Classes of compounds in which immunostimulants have been found.

Low molecular weight compounds	High molecular weight compounds
alkylamides	proteins (lectins)
phenolic compounds	peptides
alkaloids	polysaccharides
quinones	glycolipids (lectins)
saponins	
sesquiterpenes	
di- and triterpenoids	

6. References

[1] Wagner, H. and Proksch, A. (1985) Immunostimulatory Drugs of Fungi and Higher Plants, in N. Farnsworth and H. Wagner (eds.), *Economic and Medicinal Plant Research,* Academic Press (London), vol. **1**, pp 113-153.

[2] Wagner, H. (1990) Search for plant derived natural compounds with immunostimulatory activity (recent advances) *Pure & Appl. Chem.* **62**, 1217-1222.

[3] Gupta, R.K. and Siber, G.R. (1995) Adjuvants for human vaccines - current status, problems and future prospects. *Vaccine* **13**, 1263-1276.

[4] Wagner, H. and Jurcic, K. (1991) Assays for Immunomodulation and Effects on Mediators of Inflammation, in P.M. Dey, J.B. Harborne and K. Hostettmann (eds), *Methods in Plant Biochemistry*, Academic Press London - New York, vol. **6**, 195-217.

[5] Luettig, B., Steinmüller, C., Gifford, G.E., Wagner, H., Lohmann-Matthes, M.L. (1989) Macrophage activation by the polysaccharide arabinogalactan isolated from plant cell cultures of *Echinacea purpurea, J. Nat. Canc. Institute,* **81**, 669-675.

[6] Roesler, J., Steinmüller, Ch., Kiderlen, A., Emmendörfer, A., Wagner, H., Lohmann-Matthes, M.-L. (1991) Application of purified polysaccharides from cell cultures of the plant *Echinacea purpurea* to mice mediates protection against systemic infections with *Listeria* monocytogenes and *Candida albicans, Int. J. Immunopharmac.*, **13**, 27-37.

[7] Biozzi, G., Benacerraf, B., Halpern, B.N. (1953) Quantitative study of the granulopectic activity of R.E.S. II - a study of the kinetics of the granulopectic activity of the R.E.S. in relation to the dose of carbon injected. Relationship between the weight of the organs and their activity. *Brit. J. Exp. Pathol.*, **34**, 441.

[8] Brandt, I. (1967) Studies on the phagocytic activity of neutrophilic leukocytes, *Scand. J. Haematol.* (Suppl 2).

[9] Wagner, H. and Jurcic, K. (1996) A new flowcytometric assay for measuring the leukocyte phagocytosis activity of immunostimulating plant extracts, polysaccharides and various low molecular weight compounds, *Phytomedicine* **3**, (Supplement 1) p. 31.

[10] Allen, R.C. (1981), in M.D. DeLuca and W.D. McElroy (eds.) *Bioluminescence and Chemoluminescence,* Academic Press New York- London, p. 63.

[11] Meerpohl, H.G., Lohmann-Matthes, M.L., Fischer, H. (1976) Studies on the activation of mouse bone-marrow derived macrophages by the macrophage cytotoxicity factor (MCF), *Eur. J. Immunol.* **6**, 213-217.

[12] Stimpel, M., Proksch, A., Wagner, H., Lohmann-Matthes, M.L. (1984) Macrophage activation and induction of macrophage cytotoxicity by purified polysaccharide fractions from the plant *Echinacea purpurea, Infections and Immunity,* **46**, 845-849.

[13] Nakamura, S., Sung, S.S.J., Bjorndal, J.M., Fu, S.M. (1989) Human T-cell activation IV. T-cell activation and proliferation via early activation antigen EA-1, *J. Exp. Med.*, **169**, 677-689.

[14] Kabat, E.A., Mayer, M.M. (1961), in C.C. Thomas (ed.) *Kabat and Mayer´s Experimental Immunochemistry*, Springfield, Ill. U.S.A., Ed. 2, pp. 133 - 239.

[15] Platts-Mills, T.A.E., Ishizaka, K.J. (1974) Activation of the alternate pathway of human complements by rabbit cells, *J. Immunol.*, **113**, 348-357.

[16] Wagner, H., Kreher, B., Jurcic, K. (1988) In vitro stimulation of human granulocytes and lymphocytes by pico- and femtogram quantities of cytostatic agents, *Arzneim.-Forsch./Drug Res.* **38**, 237-275.

[17] Lindequist, U., Teuscher, E. (1985) Pflanzliche und mikrobielle Wirkstoffe als Immunmodulatoren, *Pharmazie*, **40**, 10-16.

[18] Eisemann, K., Totola, A., Jurcic, K., Pettit, G.R., Wagner, H. (1995), Bryostatins 1, 2 and 5 activate human granulocytes and lymphocytes: *in vitro* and *in vivo* studies. *Pharm. Pharmacol. Lett.*, **1**, 45-48.

CYCLOOXYGENASE AND 5-LIPOXYGENASE AS TARGETS FOR MEDICINAL PLANT RESEARCH

R. BAUER
Institut für Pharmazeutische Biologie der Universität Düsseldorf
Universitätsstr. 1, D-40225 Düsseldorf, Germany

Introduction

The incidence of inflammatory diseases in the field of rheumatism, allergies and asthma is increasing. 5 % of the world's population is suffering from asthma. About 10 % of the children are said to have asthmoid disorders. In the U.S.A., 5.000 people die from asthma per year. In Germany it is 1 lethal case within 100.000 inhabitants per year. In Germany, also one sixth of the population is suffering from rheumatism. Out of them, 1.2 million have chronic polyarthritis. Therefore, there is a real need for new and better anti-inflammatory drugs and concepts.

The drugs already on the market, like corticosteroids, pair their benefits with serious side effects on the mineral household and on hormonal functions. But also non steroidal anti-inflammatory compounds, like indomethacin, are burdened with side effects in the field of the blood generating and gastrointestinal system. Therefore, new leads with less side effects are desirable.

Our knowledge on the pathophysiological pathways of the inflammatory processes has been widened in recent years. A lot of mediators of inflammation are already known and also the mechanisms how these mediators are generated. The involvement of prostaglandins and leukotrienes, of the complement system and of cytokins in the process of inflammation becomes more and more understood. Accordingly, new drug research is directed to find compounds which act specificly via these mechanisms.

Involvement of cyclooxygenase and 5-lipoxygenase in the inflammatory process

Arachidonic acid is released from membrane lipids by phospholipase A_2 and is metabolized via two major enzymatic pathways: the cyclooxygenase pathway yielding the primary prostaglandins and thromboxane, and the 5-lipoxygenase pathway yielding the leukotrienes [1, 2].

Cyclooxygenase (prostaglandin endoperoxidase synthase, EC 1.14.99.1), also called prostaglandin H-synthase (PGHS), has two enzymatic properties and forms in the first step the cyclic hydroperoxide PGG_2 which is then converted by a peroxidase reaction to the corresponding alcohol PGH_2 [3]. From PGH_2 different products can be formed depending on the tissue or the cell the reaction takes place: the various prostaglandins like PGE_2, PGF_2, and PGD_2 thromboxane A_2 or prostacyclin. The activities of prostaglandins are quite complex: e.g. PGE_2 is an important mediator of pain perception, PGF_2, PGD_2 and PGG_2 induce bronchoconstriction, while PGE_2 and PGI_2 cause bronchodilatation. Therefore, COX has been regarded for a long time as the target of most nonsteroidal anti-inflammatory drugs (NSAID), like indomethacin, aspirin and piroxicam [4].

In recent years, besides the constitutive COX-1 an inducible isoenzyme COX-2 has been detected [5, 6, 7]. About 60 % homology exists between the two isoforms of COX within the species, whereas >80% homology has been observed among the inducible enzymes in different species [6, 8]. At the active site there seems to be only a difference of a single amino acid [9]. Constitutively expressed COX-1 is responsible for the homeostatic actions of prostanoids, e.g. in gastric mucosa and renal cortex, whereas inducible COX-2 is involved in the biosynthesis of eicosanoids at sites of inflammation after activation by mitogenic or inflammatory stimuli [10, 11]. However, COX-2 seems to have also importance in normal physiological processes, e.g. in ovulation and in the kidneys [5, 12, 13, 14]. COX-2 has been the target for intensive drug research in recent years [15, 16, 17, 18].

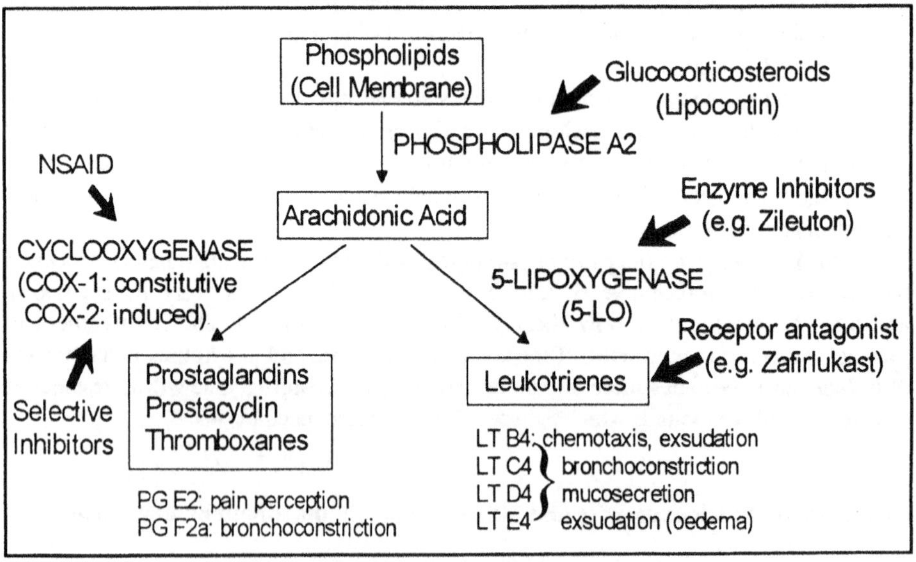

Figure 1. Arachidonic acid metabolism as a drug target.

The second pathway of arachidonic acid metabolism is leading to the leukotrienes. By **5-lipoxygenase**, arachidonic acid is transformed via 5-hydroperoxy-eicosa-tetraenoic acid (5-HPETE) into 5-HETE (5-hydroxy-eicosa-tetraenoic acid) and LTA_4. From LTA_4 the different leukotrienes are formed, e.g. LTB_4, which is a major chemotactic factor for leukocytes, or the peptidoleukotriens LTC_4, LTD_4, and LTE_4, which represent the "slow reacting substance of anaphylaxis" and react as bronchoconstricting agents. Therefore, these leukotrienes play a very important role as mediators of inflammation and asthma [19, 20, 21]. Also in case of psoriasis, allergic rhinitis, rheumatoid arthritis and colitis ulcerosa, high levels of leukotriens have been observed [22 ,23] and inhibitors of 5-LO are expected to be desirable drugs. Also compounds which block both pathways, would be of interest, in order to avoid a substrate shift [24]. They could also offer the chance to replace or diminish the use of corticosteroids, which inhibit phospholipase A_2 via lipocortin [25], but also block cytokin-induced expression of COX-2 [26].

Regarding 5-lipoxygenase, at present, three main approaches are undertaken in drug research: direct inhibition of 5-LO, antagonism of 5-lipoxygenase activating protein (FLAP), and 5-LO receptor antagonism [27, 28, 29]. Hydroxamic acid derivatives have shown to be **direct inhibitors** of 5-LO because of interaction with the non-hem Fe^{3+} in the active site of the enzyme. However, *in vivo* hydroxamic acids have been disapointing because of fast inactivation by cleavage and glucuronidation. N-hydroxy urea derivatives like zileuton are less rapidly metabolized. Zileuton (Abott: Leutrol®) has been tested in double blind study for asthma and has been found to be effective compared to placebo [30]. **FLAP antagonists** inhibit binding of 5-LO to the membrane and therefore its activation. Synthetic compounds, like the 2-quinolone derivative MK-0591 and the quinoline BAY X1005 have been found to be active via this mechanism [31, 32, 33, 34]. **LTD_4 antagonists** are already on the market as antiasthmatic drugs in Japan (Pranlukast) and Ireland (Zafirlukast) [35].

In summary, the search for inhibitors of cyclooxygenase and 5-lipoxygenase is an interesting approach for anti-inflammatory drug research and is usefull also in the screening for active compounds in plants. Appropriate *in vitro* assays on cellular, enzymatic or genetic level are available [36, 37].

Screening assays for COX and 5-LO inhibitory activity

Common COX-1 *in vitro* assays are performed using microsomal cyclooxygenase from ram [38, 39] or bovine [40, 41] seminal vesicles, or with intact cells using fibroblasts from ram seminal vesicles [42], bovine thrombocytes [43], and sheep or human platelets [44 ,45]. For the COX-2 assay, an enzyme preparation from sheep placental cotyledons can be used [46]. A human whole blood assay for COX-1 and COX-2 inhibition has previously been described [47]. But also a baculovirus expression system with infected *Spodoptera frugiperda* (sf9) cells expressing either human cyclooxygenase-1 (hCOX-1) or human cyclooxygenase-2 (hCOX-2) have been suggested for testing [48]. COX-2 is less active than COX-1, so higher enzyme levels and slightly different conditions have to be used [49]. As a substrate 1-^{14}C labelled arachidonic acid can be applied, together with reduced glutathione and epinephrine hydrogentartrate as co-substrates. In that case inhibition of prostaglandin biosynthesis can be determined specifically by HPLC separation of the radioactive metabolites and quantification of PGE_2 [50, 51]. Also a microtiter assay has been described [52]. The evaluation of the assay can also be performed by a scintillation counter after separating PGE_2, or without radioactivity by a monoclonal antibody based enzyme immuno assay, which allows a high throughput screening [53, 54, 55]. In comparison to a blank run, inhibition of prostaglandin formation can be calculated. Indomethacin is regularly used as a positive control.

For testing inhibition of formation of 5-lipoxygenase products, porcine [56, 57], bovine [43], rat [58] or human [36, 59] polymorphonuclear neutrophils (PMN) have been used. After activation of 5-LO with calcium and incubation with 1-^{14}C-arachidonic acid, inhibition of leukotriene formation can be determined by HPLC separation of 5-HETE or LTB_4 and comparison with a blank run [60 ,61 ,62]. But also methods measuring the oxygen consumption [63] and RIA [64] have been applied. Nordihydroguaiaretic acid is frequently used as a positive control.

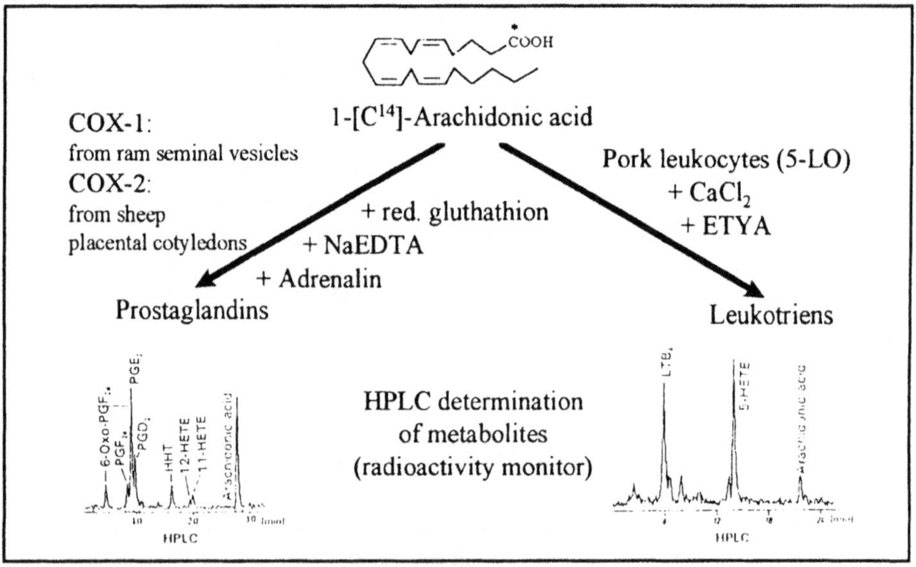

Figure .2. Cyclooxygenase and 5-lipoxygenase *in vitro* inhibition Assays.

Examples of recent research

Besides our own results and experience with COX and 5-LO testing, few extraordinary findings from other groups shall be presented. Extensive reviews on anti-inflammatory active natural products have been published elsewhere [65, 66].

Boswellia serrata

Boswellia serrata (Burseraceae) is a tree from India, which produces a resin, called "Guggulu", "Salai guggul", or "Indian frankincense". In India, an Ayurvedic preparation from the resin is on the market (Sallaki®) which is used against rheumatism and swellings of the joints. In Europe it is marketed as 'H-15".

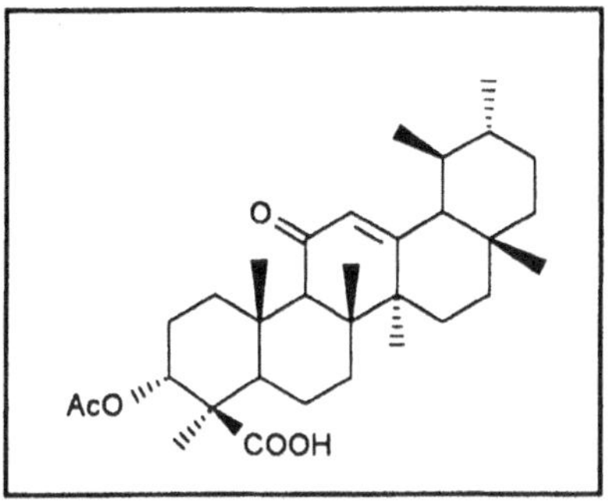

Figure 3. Acetyl-11-keto-ß-boswellic acid (AKAB).

An ethanolic extract of the gum resin of *Boswellia serrata* showed marked inhibition of LTB_4 and 5-HETE formation from endogenous and exogenously added arachidonic acid in rat peritoneal neutrophils [67]. Ammon et al. [68] found out, that boswellic acids (pentacyclic triterpenes), especially acetyl-11-keto-ß-boswellic acid (AKAB), inhibited specifically the formation of leukotrienes in human leukocytes. The IC_{50} of AKAB for 5-LO was determined to be 1.5 μM. Cyclooxygenase and lipid peroxidation were not influenced, why it has been concluded that AKAB is a specific, nonreducing-type of inhibitor which binds directly to 5-LO at a site which is different from the arachidonate binding site [69, 70]. In the meantime also clinical efficacy of extracts from the resin of *Boswellia serrata* (Salai guggul) in rheumatoid arthritis and ulcerative colitis have been reported [71, 72]. In preliminary studies, it also has been found that *Boswellia serrata* extract is effective in malignant astrocytomas, which go along with an overproduction of LTE_4 [73]. Therefore, Boswellia extracts and boswellic acids might turn out as interesting drugs in the future.

Chinese plants are an interesting source for new drug leads, because China has a long and well documented tradition and experience in using herbs medicinally. According to a calculation by But [74], from 26.092 species listed in the Flora Sinica, 4.941 (19%) are used medicinally. Out of them 1.318 species are applied in the anti-rheumatic field. Therefore, at least one fourth of the medicinal plants in China are used for anti-inflammatory purposes.

TABLE 1. Medicinally used plant species in China (TCM) with portion of anti-rheumatic use (according to But []).

	Flora R.P.Sinicae	TCM	Anti-rheumatic
Species (vasc. plants)	26.092	4.941 (19%)	1.318
Genus	2.949	1.544 (52%)	655
Families	287	240 (84%)	161

Houttuynia cordata

After a selection based on ethnomedical data, we have performed a pharmacological screening with 80 plants. The most active ones were considered for further investigations. *Houttuynia cordata* (Saururaceae) was highly active in terms of COX-1 inhibition. The aerial parts of this plant have been used in traditional Chinese medicine as a "detoxicant", anti-inflammatory and anti-pyretic agent. By activity guided fractionation we have tried to find the active constituents regardless to their chemical nature.

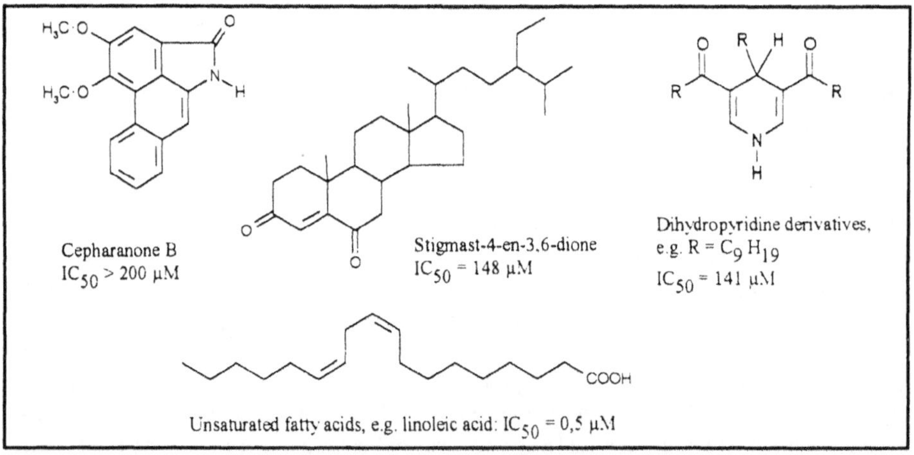

Figure 4. Constituents from the *n*-hexane extract of *Houttuynia cordata* and their COX-1 inhibitory activity.

Oleic acid and linoleic acid turned out to be the most active compounds and responsible for the activity of the extract. They exhibited IC_{50} values of 13,4 and 0,25 μM respectively (indomethacin: IC_{50} = 1.4 μM) [75]. The anti-inflammatory activity of unsaturated fatty acids is well documented. They are metabolized by COX in a similar way as arachidonic acid [76] to compounds like PGE_1 (derived from dihomo-GLA) which have an anti-inflammatory effect. This is the reason, why oils and plants with a high content of unsaturated fatty acids, like evening primrose, borage seeds, or fish oils are used in inflammatory diseases [77]. Since the concentration of unsaturated fatty acids in *H. cordata* is 0.3 - 1.1%, they might exert activity when the plant is eaten as a vegetable diet. It is also known, that linoleic acid is metabolized by 5-LO to 9(S)hydroxyperoxy-10E,12Z-octadecadienoic acid (9-HPODE) so that it obviously competes with arachidonic acid also for this enzyme [78, 79]. The other compounds which were isolated from *H. cordata* were less active. Phytol showed an IC_{50} value of 44 μM, and stigmast-4-ene-3,6-dione of 148 μM. Moreover 6 pyridine and dihydropyridine derived alkaloids could be identified, which represented new natural products [80]. Their basic structures are found in well known synthetic calcium channel blocking agents like nifedipin. However, the compounds from *H. cordata* showed no Ca-antagonistic effect. Therefore, their pharmacological potential has still to be determined.

Atractylodes lancea

The rhizomes of *Atractylodes lancea* (Compositae) are used in China to treat rheumatic diseases, digestive disorders, mild diarrhea and influenza. They are rich in sesquiterpenes and acetylenic compounds such as atractylodin, atractylodinol and acetylatractylodinol [81], which dominate also in the *n*-hexane extract. The *n*-hexane extract exhibited very good inhibitory activity *in vitro* of 5-LO (IC_{50} = 2,9 μg/ml) and moderate inhibition of COX-1 (IC_{50} = 30,5 μg/ml) [82]. However, none of the major compounds was responsible for that strong inhibitory effect. Consequent activity guided fractionation of the extract led to the detection of highly active minor constituents: a new chromene (atractylochromene) with IC_{50} values of 0,56 μM (5-LO) and 3,3 μM (COX-1), a quinone, (2'*E*)-2-(3',7'-dimethylocta-2',6'-dienyl)- 6-methyl-2,5- cyclohexadiene-1,4-dione (atractyloquinone) - known already from *Atractylis koreana* [83] - with an even stronger, selective inhibitory activity on 5-LO (IC_{50} (5-LO) = 0,23 μM, IC_{50} (COX-1) = 64,3 μM), and the coumarin osthol, which was isolated for the first time from this genus (IC_{50} (5-LO) = 27,9 μM) [84]. The chromene and quinone represent highly active and selective inhibitors of 5-LO. Similar quinones have previously been isolated only from brown algae [85]. The active compounds were also tested for influencing the *ex vivo* production of oxygen radicals by human polymorphonuclear cells (PMN) stimulated with FMLP or opsonized zymosan (OZ). Furthermore their free radical scavenging activity was studied in a cell-free horseradish peroxidase *in vitro* system. The determined IC_{50} values were 1,3 μM (PMN, FMLP) and 5,6 μM (PMN, OZ) for atractylochromene and 1,1 μM (PMN, FMLP) and 5,4 μM (PMN, OZ) for atractyloquinone. Therefore, both exhibited remarkable inhibitory effect in all *ex vivo* test systems, but they showed much less activity in the horseradish peroxidase *in vitro* system. So, it seems that they may need to be activated in vivo before exerting antioxidative properties [86].

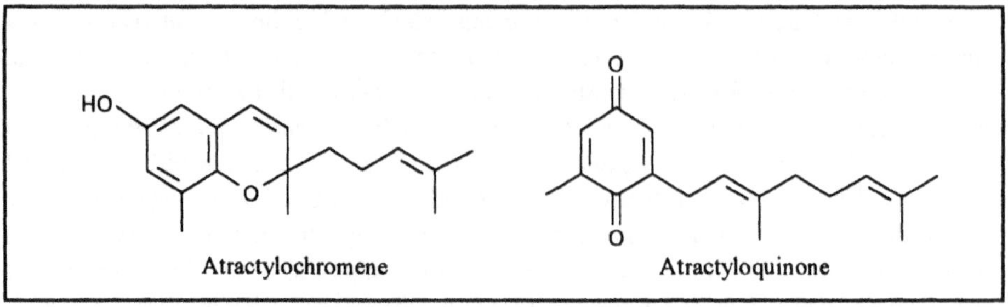

Figure 5. 5-LO and COX-1 inhibitory active constituents from *Atractylodes lancea* rhizomes.

Bidens campylotheca

By the according strategy, the active compounds were detected in the herb of *Bidens campylotheca* (Compositae), which has been used on the Hawaiian islands under the name "Ko´oko´olau" for strengthening the body ("tonic"), to treat throat and stomach trouble, to stimulate appetite, and for bad cases of asthma. The COX-1 inhibitory active *n*-hexane extract (IC_{50} = 8 μg/ml) consisted mainly of polyacetylenes, which were identified as unsaturated derivatives of heptadecane, safynol and safynol-isobutyrate respectively.

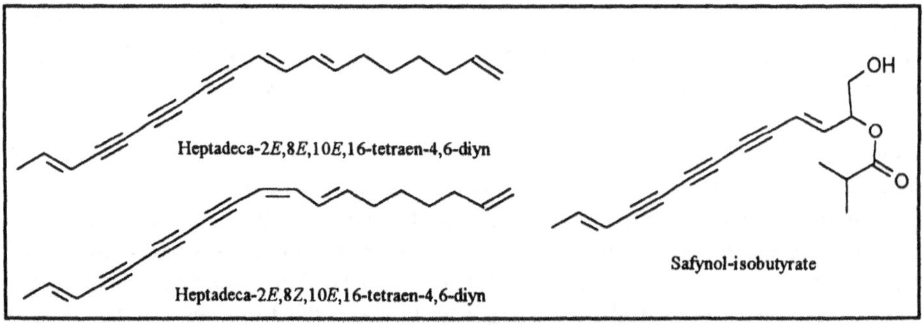

Figure 6. Polyacetylenes from *Bidens campylotheca.*

The constituents have been tested for COX-1 and 5-LO inhibitory activity and it turned out that they represent an active principle of the extract, showing IC_{50}-values of 10 - 180 μM on COX-1. The 8Z-isomer of heptadeca-2E,8,10E,16-tetraen-4,6-diyn was more potent than the trans isomer (IC_{50} = 110 μM versus 180 μM). Therefore, a strong structure-activity relationship is evident. Safynol-isobutyrate, which is a new natural product, appeared to be the most active compound (IC_{50} = 10 μM). Its actvity was almost as high as the activity of indomethacin. Safynol, the corresponding alcohol, had much less activity than the ester (IC_{50} = 185 μM), which is in analogy to the activity of aspirin and salicylic acid. Safynol-isobutyrate also inhibited 5-LO in a very effective way (100 % inhibition at 35 μM) and can therefore be regarded as a "dual inhibitor" of COX-1 and 5-LO [52]. Recently, Morita et al. [87] reported on polyacetylenes from *Bupleurum falcatum*, which also exhibited a dual inhibitory effect on COX and 5-LO. Also panaxynol, a polyacetylene from *Panax ginseng*, showed inhibitory effects on mammalian lipoxygenases [88]. Therefore, polyacetylenes seem to be potent anti-inflammatory compounds.

Echinacea

Alkamides, N-containing compounds related to polyacetylenes, have been found to possess immunomodulatory activity and to represent one of the active constituens of *Echinacea* species [89]. The lipophilic fraction of *E. angustifolia* roots consists of at least 15 different polyunsaturated alkamides [90] and has shown inhibitory effects on both COX-1 and 5-LO [91].

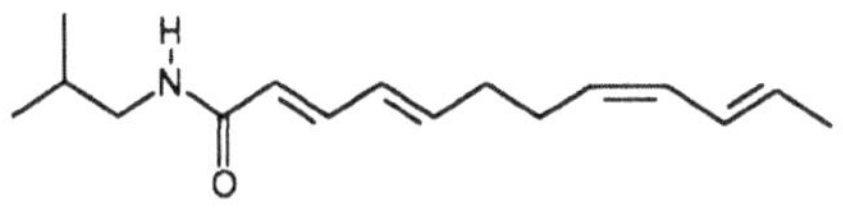

Figure 7. Dodeca-2*E*,4*E*,8*Z*,10*E*-tetraenoic acid isobutylamide, the major alkamide from *Echinacea angustifolia* and *E. purpurea*.

Testing the individual alkamides, it turned out, that the long chain compounds, pentadeca-2E,9Z-dien-12,14-diynoic acid isobutylamide and hexadeca-2E,9Z-dien-12,14-diynoic acid isobutylamide showed the strongest inhibitory effect on COX, while inhibition of 5-LO was quite weak (see Fig. 8). Several compounds showed no effect on 5-LO at all. However, dodeca-2,4,8,10-tetraenoic acid isobutylamide, the main constituent of the roots showed inhibitory activity mainly on 5-LO (IC_{50} = 40 μM) and hardly on COX-1 (IC_{50} = 202 μM) [92]. Therefore, the alkamides represent an important anti-inflammatory active principle of *Echinacea angustifolia* and might explain the traditional use of *Echinacea* roots against tooth ache. As some alkamides inhibit leukotriene biosynthesis, they might be also responsible for the antiallergic potential of *Echinacea* extracts. Due to structural similarity to leukotrienes, it can be argued that they can act also as leukotriene antagonists.

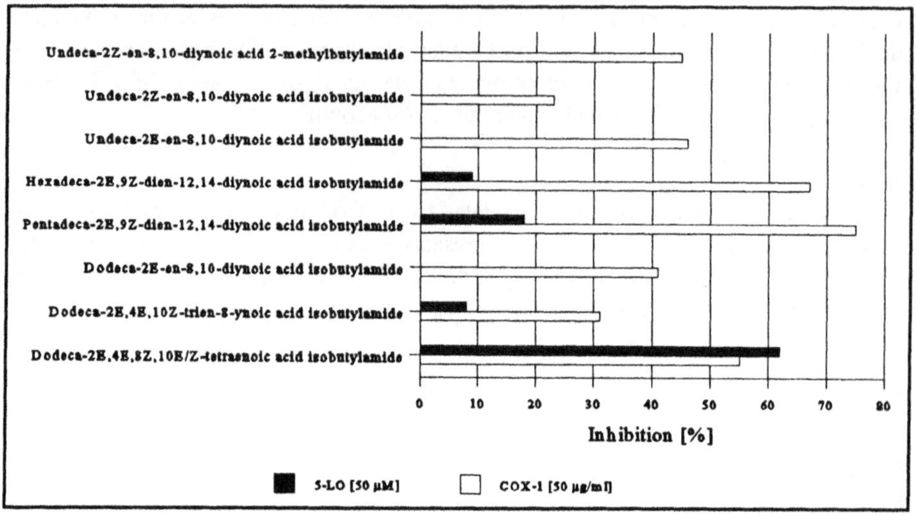

Figure 8. Inhibition of prostaglandin and leukotriene biosynthesis by alkamides from *Echinacea angustifolia.*

Zanthoxylum chalybeum

Zanthoxylum chalybeum (Rutaceae) is a tree which grows in Africa. Its root bark is used in Tanzania for the treatment of rheumatism, swellings, pains, and pneumonia. The dichloromethane extract of the root bark showed marked inhibition of 5-LO with an IC_{50} value of 7 μg/ml. The *n*-hexane extract exhibited less activity (IC_{50} = 23 μg/ml) and the methanolic extract was almost inactive. The effects on COX-1 were less pronounced. The known benzophenanthridine alkaloids chelerythrine and nitidine, as well as the protoberberine bases jatrorhizine and palmatine could be identified as the major constituents of the DCM extract [93, 94, 95]. The protoberberine alkaloids hardly inhibited 5-LO and had no effect on COX-1. However, benzophenanthridine alkaloids showed a significant and selective activity on 5-LO. From a series of benzophenanthridine alkaloids chelerythrine and sanguinarine turned out as the most potent inhibitors of 5-LO with IC_{50} values of 2,0 and 2,7 μM respectively [95,96]. Since both constituents did not show inhibition of peroxidation, it is likely, that the iminium ion of benzophenanthridin alkaloids, which can undergo disproportion to 5,6-dihydro and 6-oxo-derivatives, forms nucleophilic adducts with SH-groups of the enzyme or of FLAP [97]. A similar activity has also been found for extracts and constituents of *Chelidonium majus* [98]. Chelidonine, sanguinarine, and chelerythrine also inhibit taxol-mediated polymerization of rat brain tubulin in the micromolar range [99] and possess a potent inhibitory action against the growth of human keratinocytes [100]. Therefore they might be good candidates for further drug development.

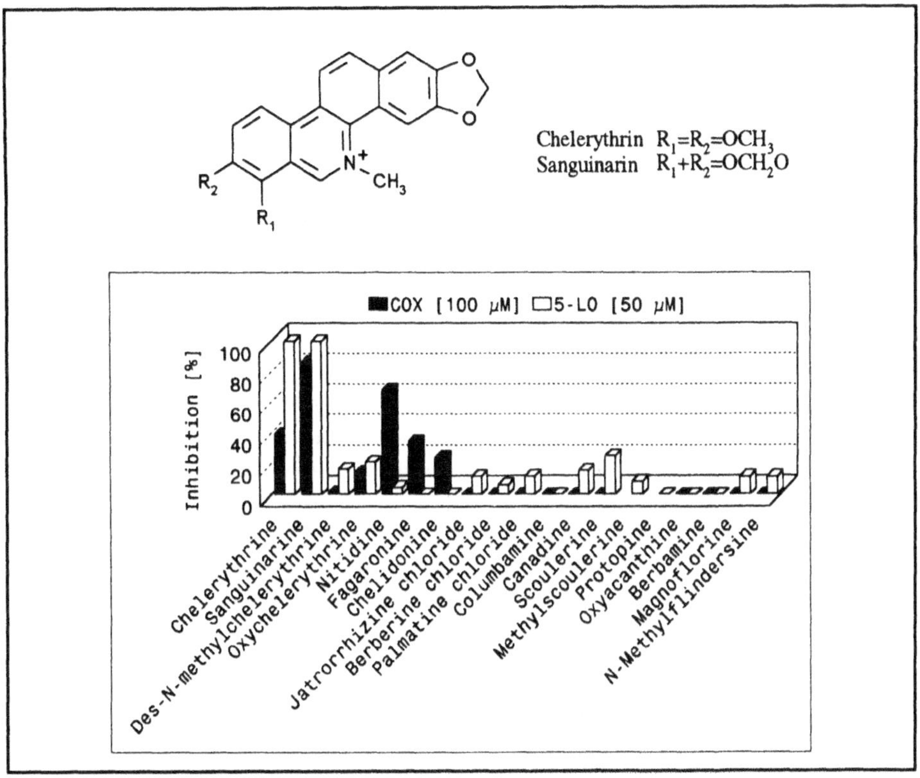

Figure 9. In vitro inhibition of 5-LO and COX by benzophenanthridine alkaloids.

Zanthoxylum armatum

Zanthoxylum armatum (Rutaceae) is a common plant in Southeast Asia and is used in Nepal and China as a tonic in case of fever, against cholera and for tooth ache. Hexane and DCM extracts of the bark of *Z. armatum* showed significant inhibitory effect on 5-lipoxygenase (IC_{50} 6,4 and 6.1 µg/ml respectively). Prostaglandin synthase inhibitory activity was less pronounced. Both in the *n*-hexane and the DCM extract, several furofurano lignans could be identified [95, 101, 102, 103]. The major constituents were eudesmin, magnolin, epieudesmin, kobusin, fargesin, planinin, and sesamin.

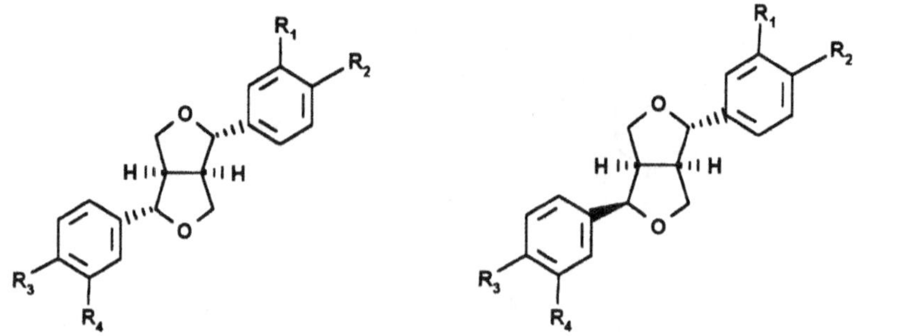

(+)-Sesamin	$R_1/R_2 = OCH_2O$, $R_3/R_4 = OCH_2O$	(+)-Asarinin	$R_1/R_2 = OCH_2O$, $R_3/R_4 = OCH_2O$
(+)-Kobusin	$R_1=R_2 = OCH_3$, $R_3/R_4 = OCH_2O$	(+)-Fargesin	$R_1=R_2 = OCH_3$, $R_3/R_4 = OCH_2O$
Eudesmin	$R_1=R_2=R_3=R_4 = OCH_3$	Planinin	$R_1/R_2 = OCH_2O$, $R_3=R_4 = OCH_3$
		(-)-Phylligenin	$R_1= OCH_3$, $R_2 = OH$, $R_3=R_4 = OCH_3$
		Epieudesmin	$R_1=R_2=R_3=R_4 = OCH_3$

Figure 10. Tetrahydro-furofurano lignans from *Zanthoxylum armatum.*

When testing the compounds for COX-1 and 5-LO inhibitory activity, it became obvious that sesamin was the main active compound with an IC_{50} value of 17 μM on 5-LO (see Fig. 9). Asarinin, which is the epimer of sesamin, was less active. Also compounds lacking one or two methylene dioxo functions, like kobusin and eudesmin were less active. The reference inhibitor nordihydroguaiaretic acid, however, which is a diarylbutanolignan and a well known 5-lipoxygenase inhibitor, was 10 times more potent (IC_{50} = 1,5 μM) [104]. COX-1 inhibitory activity of most of the lignans was not significant. Only fargesin (100 μM: 47 % inhibition) and planinin (100 μM: 30 % inhibition) showed weak effects. Therefore, the majority of the lignans are selective 5-LO inhibitors. In order to establish more data on structure activity relationship, we have tested further tetrahydrofurofuran lignans from other plants [105, 106] and from synthetic origin [107]. (±)-Pinoresinol turned out as a potent 5-LO inhibitor, but also inhibited COX-1. (+)-Episesartemin B, (+)-epiyangambin and (±)-paulownin exhibited no inhibitory activity. (+)-Epiashantin, (±)-syringaresinol, (-)-prenylpiperitol und (-)-prenylpluviatol were identified as weak inhibitors of 5-LO. (-)-Prenylpiperitol also showed COX-1 inhibitory activity [104].

Structure activity relationship of tetrahydrofurofurano lignans can be summarized as follows [104]:

- Diequatorial substituted tetrahydrofurofuran lignans are more inhibitory active on 5-LO than endo-exo substituted derivatives.
- Piperonyl moieties turned out as superior compared to veratryl groups.
- Methoxy groups had negative effects on activity, while methylenedioxy functions were enhancing activity.
- Hydroxy substituents at the bridge carbons, like in (±)-paulownin, diminished activity completely.

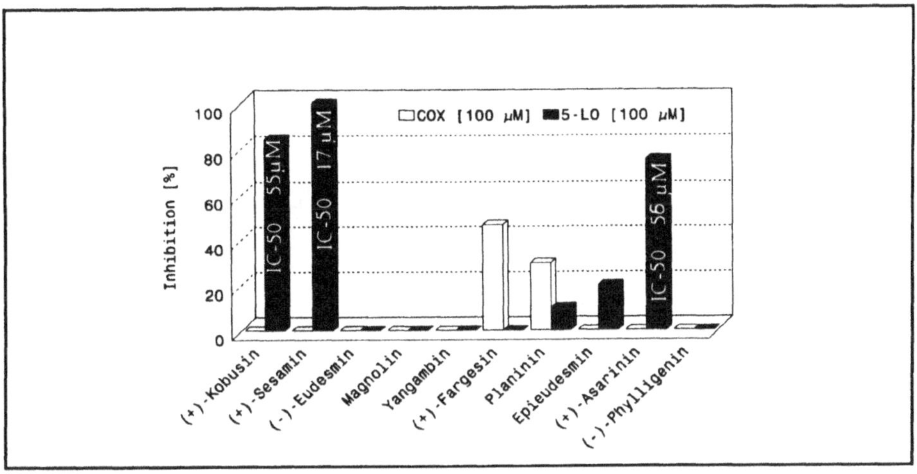

Figure 11. COX-1 and 5-LO inhibitory activity of tetrahydro-furofurano lignans from the bark of Zanthoxylum armatum.

The active compounds were also tested for antioxidative properties in the peroxidation assay, but showed no effect. Therefore, 5-lipoxygenase inhibition is not achieved because of radical scavenging activity. It can be argued, that inhibition is probably accomplished via a reversible "dead end" complex with 5-LO as suggested by Hutchinson *et al.* [108]. Sesamin is regarded as an interesting lead for further development. It has also shown inhibitory effect on delta-5-desaturase and therefore also interacts with the cholesterol biosynthesis [109].

The methoxylated lignans (-)-phylligenin, (+)-kobusin, (-)-eudesmin and (±)-epieudesmin turned out as potent inhibitors of PAF induced thrombocyte aggregation as it is known for many other lignans [110, 111]. The IC_{50} values were less than 50 μM. (+)-Sesamin and (+)-asarinin however, exhibited no activity [104]. This again shows, that 5-LO inhibition of sesamin is a very specific effect and worthwhile for further investigation.

In the pericarps of *Zanthoxylum armatum*, 5-LO inhibitory activity could be attributed to the methoxylated flavonoid tambulin (IC_{50} = 28,2 μM) [104]. Flavonoids are well known as inhibitors of 5-LO [112, 113]. Also quercetin has shown to be very effective (IC_{50} = 3 μM) [62]. They have also shown activity *in vivo* [114]. Flavonoids act by chelating and reducing the central ferrum atom of the enzyme. The alkamides hydroxy- α-sanshooil and hydroxy-β -sanshooil, which are also major constituents of the pericarps, exhibited only weak or medium activity on 5-LO and almost no activity on COX-1 (see Fig. 12). β-Sitosterol contributes to the COX-1 inhibiting properties of the extract (IC_{50} = 23 μM). So, it is evident that the activity of the plant is derived from several compounds. It is a question whether all act additively or synergisticly.

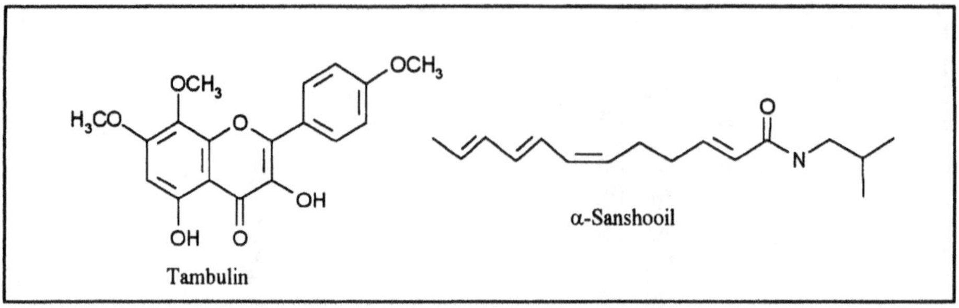

Figure 12. COX-1 and 5-LO inhibitory active constituents from the pericarps of *Zanthoxylum armatum*.

When testing α-sanshooil, an alkamide with dodeca-2E,6Z,8E,10E-tetraenoic acid structure and sesamin together in mixtures applying the isobol method [115, 116], we could observe a synergistic effect (see Fig. 13). The straight line represents additive effects without interaction. When the IC_{50} of the mixture is lower, it is significant for synergistic action. As Fig. 14 demonstrates, (+)-sesamin in combination with the alkamide -sanshooil showed synergistic inhibitory activity on 5-LO [104]. This may be an explanation of better activity of plant extracts compared to pure compounds. And it shows, that the activity of *Zanthoxylum armatum* is probably derived from the tetrahydrofurofurano lignans and the alkamides acting together.

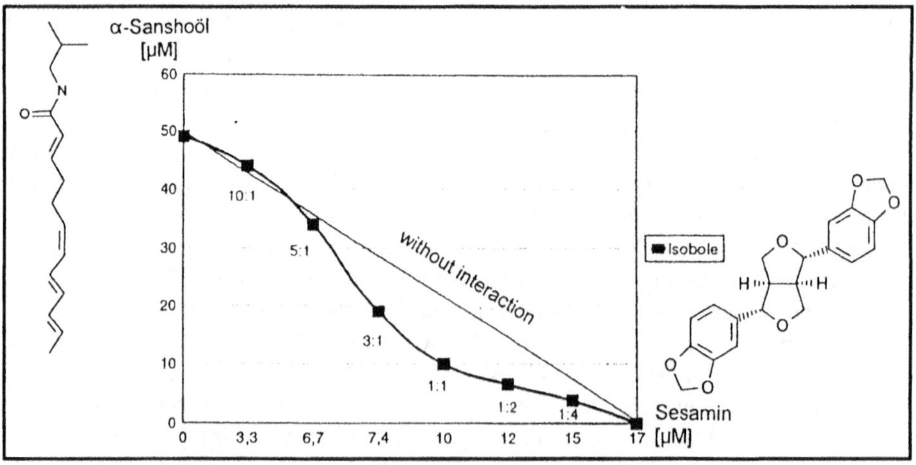

Figure 13. Synergistic effect of α-sanshooil and sesamin on 5-lipoxygenase determined with the isobol method.

Angelica pubescens

Duhuo, the roots of *Angelica pubescens* Maxim *f. biserrata* (Apiaceae) have been used in traditional Chinese medicine as a remedy for arthritic disease. More than thirty coumarins and other compounds have been reported as constituents [117, 118, 119, 120]. Linoleic acid, osthol, osthenol and two polyacetylenes, falcarindiol and 11(S),16(R)-dihydroxy-octadeca-9Z,17-dien-12,14-diyn-1-yl acetate were found to be the most active compounds responsible for the inhibitory activity of the dichloromethane extract of the roots of *Angelica pubescens* on 5-LO and COX-1 *in vitro*. They showed prominent inhibitory effect on 5-LO with IC_{50} values of 27.9 μM, 36.2 μM, 43.1 μM, 9.4 μM and 24.0 μM respectively. Linoleic acid, osthenol, falcarindiol and 11(S),16(R)-dihydroxy-octadeca-9Z,17-dien-12,14-diyn-1-yl acetate exhibited inhibitory activity on COX-1 with IC_{50} values of 13.3 μM, 64.3 μM, 66.0 μM and 73.3 μM [121].

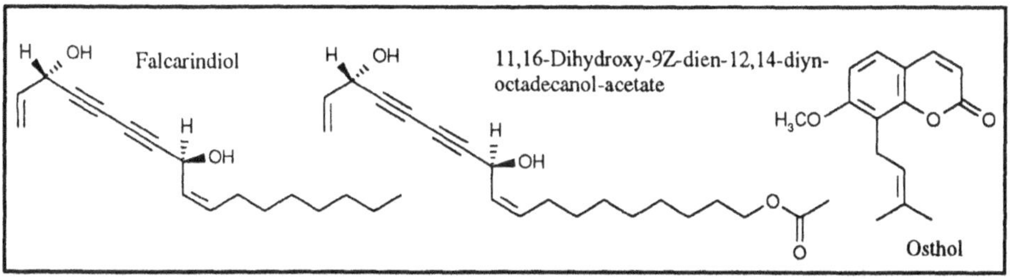

Figure 14. 5-LO and COX-1 inhibitory active constituents from the roots of *Angelica pubescens.*

Notopterygium incisum

The Chinese drug "*Qianghuo*" represents the underground parts of *Notopterygium incisum* Ting ex H.T.Chang (Apiaceae). They are frequently used in TCM against headache, common cold, and rheumatic disorders. The *n*-hexane extract of *Qianghuo* showed inhibitory activity *in vitro* of 5-LO and COX-1. By bioactivity guided isolation phenethyl ferulate, falcarindiol, and as a minor compound (-)-bornyl ferulate were identified as the main active principles. Phenethyl ferulate represents a dual inhibitor of COX-1 (IC_{50} = 4,35 μM) and 5-LO (IC_{50} = 5,75 μM). Also (-)-bornyl ferulate exhibited remarkably low IC_{50} values of 10,4 μM (5-LO) and 12 μM (COX-1). Free ferulic acid did not show any inhibitory effect neither on COX nor on 5-LO up to the concentration of 100 μM. Therefore, the activity seems not to be derived from the methoxy-phenol moiety. Testing the antioxidant properties in the peroxidation assay, phenethyl ferulate and (-)-bornyl ferulate showed only a moderate effect with IC_{50} values of 47 μM and 62 μM respectively. So, 5-LO and COX inhibitory activity of phenethyl ferulate and (-)-bornyl ferulate is not based on their radical scavenging properties [134].

Falcarindiol turned out as potent inhibitor of 5-LO (IC_{50} = 9,4 μM) and only a weak inhibitor of COX-1 (IC_{50} = 66 μM) and therefore it almost selectively inhibits 5-LO. In the concentration employed in the 5-LO assay (below 15 μM) it exhibited no evident cytotoxic effect. In the peroxidation assay, falcarindiol showed a strong antioxidant activity (IC_{50} = 6,85 μM), which can be one explanation of its inhibitory effect on 5-LO. Falcarindiol may also be responsible for the antioxidant potency of the methanolic extract of *N. incisum* described by Yang *et al.* [122]. The furanocoumarins isoimperatorin and notopterol hardly showed any inhibitory effect neither on COX-1 nor on 5-LO. In previous investigations, notopterol has been identified as an antiinflammatory active principle of *N. incisum* because of effects in the acetic acid-writhing test in mice [122]. We found that notopterol exhibits only weak inhibitory activity on 5-LO (IC_{50} = 47,5 μM) and had no activity on COX-1 [134].

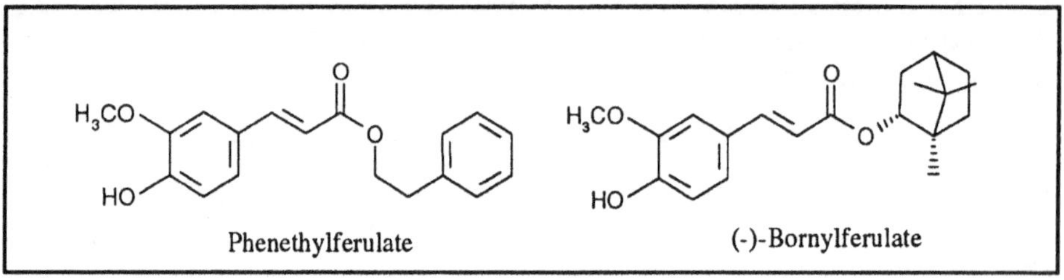

Figure 15. 5-LO and COX-1 inhibitory active constituents from the roots of *Notopterygium incisum*.

In a recent investigation, synergistic effects of falcarindiol and phenethyl ferulate in the inhibition of 5-LO could also be demonstrated [123]. Phenethylferulate was shown to inhibit COX-2 (IC_{50} = 2,8 μM) in a similar extent as COX-1 [124]. It is interesting that a related compound, phenethyl caffeate from propolis was shown to be both, a potent inhibitor of 5-LO [125] and a specific inhibitor of the activation of the nuclear transcription factor NF- B [126]. Transcription factors like NF-kB and NF-IL6 play an essential role in the expression of cyclooxygenase-2, but may also act independently [127 ,128 ,129]. So it would be interesting to test, whether phenethyl ferulate has a similar effect.

Recently it could be demonstrated that sesquiterpene lactones, such as helenalin, could also inhibit activation of transcription factor NF-kB due to a modification of the NF-kB/I-kB complex preventing the release of I-kB [130]. This could give a molecular mechanism for the anti-inflammatory activity of sesquiterpene lactones and the efficacy of herbal drugs like *Arnica montana*.

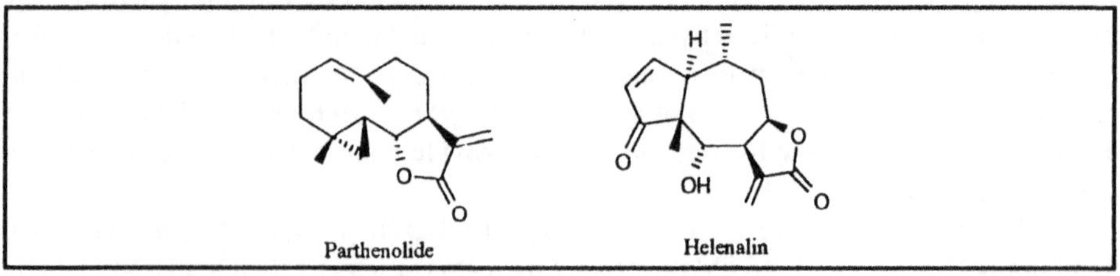

Figure 16. Parthenolide, an inhibitor of expression of COX-2 and helenalin, an inhibitor of NF-kB.

The sesquiterpene lactone parthenolide, a major constituent of *Tanacetum parthenium* (European feverfew), recently was also found to inhibit the expression of COX-2 and proinflammatory cytokins (TNFα and IL-1) in lipopolysaccharide stimulated macro-phages [131]. This effect was correlated with the suppression of LPS-stimulated protein-tyrosine phosphorylation, mainly of MAPKs (mitogen-activated protein kinases). The -methylene-gamma-lactone moiety obviously confers this inhibitory effects. This offers the possibility that some sesquiterpene lactones might be used for the treatment of septic shock, caused by LPS stimulating protein tyrosin phosphorylation of MAPKs in macrophages [131]. Therefore, inhibition of expression of COX-2 seems also to be an interesting target for future drug research [132, 133].

In summary, arachidonic acid metabolism and especially cyclooxygenase and 5-lipoxygenase as well as their activation process, are interesting targets for drug research. They are usefull also for the screening of natural products, as several examples have demonstrated. Since the plant kingdom is still far from being fully explored, there is still the chance to find new and interesting leads for future drug development [134].

References

1 Samuelsson, B.: Leukotrienes and other lipoxygenase products, *Prog. Lipid Res.* **25** (1986), 13-18.

2 Vane, J.R. and Botting, R.M.: Inflammation and the mechanism of action of anti-inflammatory drugs, *FASEB J.* **1** (1987), 89–96.

3 Smith, W.L. and Marnett, L.J.: Prostaglandin endoperoxide synthase: structure and catalysis, *Biochim. Biophys. Acta* **1083** (1991), 1-17.

4 Vane, J.R. and Botting, R.M.: The mode of action of anti-inflammatory drugs, *Postgrad. Med. J.* **66** Suppl. 4, (1990), S2-S17.

5 Sirois, J. and Richards, J.S.: Purification and characterization of a novel, distinct isoform of prostaglandin endoperoxide synthase induced by human chorionic gonadotropin in granulosa cells of rat preovulatory follicles, *J. Biol. Chem.* **267** (1992), 6382-6388.

6 Hla, T. and Neilson, K.: Human cyclooxygenase-2 cDNA. *Proc. Natl. Acad. Sci USA* **89** (1992), 7384-7388.

7 Herschmann, H.R.: Prostaglandin synthase 2, *Biochim Biophys Acta* **1299** (1996), 125-140.

8 Xie, W., Robertson, D.L., and Simmons, D.L.: Mitogen-inducible prostaglandin G/H synthase: A novel target for nonsteroidal antiinflammatory drugs, *Drug Dev. Res.* **25** (1992), 249-265.

9 Gierse, J.K., McDonald, J.J., Hauser, S.D., Rangwala, S.H., Koboldt, C.M., and Seibert, K.: A single amino acid difference between cyclooxygenase-1 (COX-1) and -2 (COX-2) reverses the selectivity of COX-2 specific inhibitors, *J. Biol. Chem.* **271** (1996), 15810-15814.

10 Herrmann, F., Lindemann, A., Gauss, J., and Mertelsmann, R.: Cytokine-stimulation of prostaglandin synthesis from endogenous and exogenous arachidonic acids in polymor-phonuclear leukocytes involving activation and new synthesis of cyclooxygenase. *Eur. J. Immunol.* **20** (1990), 2513-2516.

11 Funk, C.D., Funk, L.B., Kennedy, M.E., Pong, A.S., and Fitzgerald, G.A.: Human platelet/erythroleukemia cell prostaglandin G/H synthase: cDNA cloning, expression, and gene chromosomal assignment, *FASEB J.* **5** (1991), 2304-2312.

12 Harris, R.C., McKanna, J.A., Akai, Y., Jacobson, H.R., Dubois, R.N., and Breyer, M.D.: Cyclooxygenase-2 is associated with the macula densa of rat kidney and increases with salt restriction, *J. Clin. Invest.* **94** (1994), 2504-2510.

13 Pugliese, F. and Cinotti, G.A.: Nonsteroidal anti-inflammatory drugs (NSAIDs) and the kidney, *Nephrol. Dial. Transplant.* **12** (1997), 386-388.

14 Komhoff, M., Grone, H.J., Klein, T., Seyberth, H.W., and Nusing, R.M.: Localization of cyclooxygenase-1 and -2 in adult and fetal human kidney: implication for renal function, *Am. J. Physiol.* **272** (1997), F460-468.

15 Vane, J.R. and Botting, R.M.: New insights into the mode of action of anti-inflammatory drugs, *Inflamm. Res.* **44** (1995), 1-10.

16 Klein, T., Nüsing, R.M., Pfeilschifter, J., and Ullrich, V.: Selective inhibition of cyclooxygenase 2. *Biochem. Pharmacol.* **48** (1994), 1605–1610.

17 Mitchell, J.A., Akarasereenont, P., Thiemermann, C., Flower, R.J., and Vane, J.R.: Selectivity of nonsteroidal antiinflammatory drugs as inhibitors of constitutive and inducible cyclooxygenase, *Proc. Natl. Acad. Sci. USA* **90** (1993), 11693-11697.

18 Copeland, RA., Williams, JM., Giannaras, J., Nurnberg, S., Covington, M., Pinto, D., Pick, S., and Trzaskos, JM.: Mechanism of selective inhibition of the inducible isoform of prostaglandin G/H synthase, *Proc. Natl. Acad. Sci. USA* **91** (1994), 11202-11206.

19 Pauwels, R.A., Joos, G.F., and Kips, J.C.: Leukotrienes as therapeutic target for asthma, *Allergy* **50** (1995), 615-622.

20 Ford-Hutchinson, A.W.: Regulation of leukotriene biosynthesis, *Canc. Metastasis Rev.* **13** (1994), 257-267.

21 Wenzel, S.E.: Arachidonic acid metabolites: mediators of inflammation in asthma, *Pharmacotherapy.* **17** (1997), 3S-12S.

22 Ford-Hutchinson, A.W.: Leukotrienes B_4 in inflammation, *Crit. Rev. Immunol.* **10** (1990), 1-12.

23 Mayatepek, E. and Hoffmann, G.F.: Leukotrienes: biosynthesis, metabolism, and pathophysiologic significance, *Pediatr. Res.* **37** (1995), 1-9.

24 Laufer, S.A., Augustin, J., Dannhardt, G., and Kiefer, W.: (6,7-Diaryldihydropyrrolizin-5-yl)acetic acids, a novel class of potent dual inhibitors of both cyclooxygenase and 5-lipoxygenase, *J. Med. Chem.* **37** (1994), 1894-1897.

25 Davidson, F.F., Dennis, E.A., Powell, M., and Glenney, J.R.jr: Inhibition of phospholipase A2 by "lipocortins" and calpactins. An effect of binding to substrate phospholipids, *J. Biol. Chem.* **262** (1987), 1698-1705.

26 Kujubu, D.A. and Herschman, H.R.: Dexamethasone inhibits mitogen induction of the TIS10 prostaglandin synthase/cyclooxygenase gene, *J. Biol. Chem.* **267** (1992), 7991-7994.

27 Mantri, P. and Wittiak, D.T.: Inhibitors of cyclooxygenase and 5-lipoxygenase, *Curr. Med. Chem.* **1** (1994), 328-355.

28 Smith, L.J.: Leukotrienes in asthma. The potential therapeutic role of antileukotriene agents, *Arch. Intern. Med.* **156** (1996), 2181-2189.

29 Harris, R.R., Carter, G.W., Bell, R.L., Moore, J.L., and Brooks, D.W.: Clinical activity of leukotriene inhibitors, *Int. J. Immunopharmacol.* **17** (1995), 147-156.

30 Wenzel, S.E. and Kamada, A.K.: Zileuton: the first 5-lipoxygenase inhibitor for the treatment of asthma, *Ann. Pharmacother.* **30** (1996), 858-864.

31 Ford-Hutchinson, A.W.: FLAP: a novel drug target for inhibiting the synthesis of leukotrienes, *Trends Pharmacol. Sci.* **12** (1991), 68-70.

32 Uematsu, T., Kanamaru, M., Kosuge, K., Hara, K., Uchiyama, N., Takenaga, N., Tanaka, W., Friedman, BS., and Nakashima, M.: Pharmacokinetic and pharmacodynamic analysis of a novel leukotriene biosynthesis inhibitor, MK-0591, in healthy volunteers, *Br. J. Clin. Pharmacol.* **40** (1995), 59-66.

33 Hatzelmann, A., Goossens, J., Fruchtmann, R., Mohrs, K.H., Raddatz, S., and Müller-Peddinghaus, R.: Inversely-correlated inhibition of human 5-lipoxygenase activity by BAY X1005 and other quinoline derivatives in intact cells and a cell-free system implications for the function of 5-lipoxygenase activating protein, *Biochem. Pharmacol.* **47** (1994), 2259-2268.

34 Mancini, J.A., Prasit, P., Coppolino, M.G., Charleson, P., Leger, S., Evans, J.F., Gillard, J.W., and Vickers, P.J.: 5-Lipoxygenase-activating protein is the target of a novel hybrid of two classes of leukotriene bioxynthesis inhibitors, *Molecular Pharmacol.* **41** (1992), 267-272.

35 Sorkness, C.A.: The use of 5-lipoxygenase inhibitors and leukotriene receptor antagonists in the treatment of chronic asthma, *Pharmacotherapy.* **17** (1997), 50S-54S.

36 Weithmann, K.U.: Constitutive and inducible cellular arachidonic acid metabolism in vitro, in: H.G. Vogel and W.H. Vogel: *Drug Discovery and Evaluation*, Springer Verlag, Berlin, Heidelberg, 1997.

37 Harvey, J. and Osborne, D.J.: A rapid method for detecting inhibitors of both cyclo-oxygenase and lipoxygenase metabolites of arachidonic acid, *J. Pharmacol. Meth.* **9** (1983), 147-155.

38 Van der Ouderaa, F.J. and Buytenhek, M.: Purification of PGH synthase from sheep vesicular glands, *Methods Enzymol.* **86** (1982), 60-68.

39 Van der Ouderaa, F.J., Buytenhek, M., Nugteren, D.H., and Van Dorp, D.A.: Purification and characterisation of prostaglandin endoperoxide synthetase from sheep vesicular glands, *Biochim. Biophys. Acta* **487** (1977), 315-331.

40 White, H.L. and Glassman, A.T.: A simple radiochemical assay for prostaglandin synthetase, *Prostaglandins* **7** (1974) 123-129.

41 Pongprayoon, U., Baeckstrom, P., Jacobsson, U., Lindstrom, M., and Bohlin, L.: Compounds inhibiting prostaglandin synthesis isolated from *Ipomoea pes-caprae, Planta Med.* **57** (1991), 515-518.

42 Freyberger, A., Schnitzler, R., Schiffmann, D., and Degen, G.H.: Prostaglandin-H-synthase competent cells derived from ram seminal vesicles: a tool for studying cooxidation of xenobiotics, *Mol. Toxicol.* **1** (1987-88), 503-512.

43 Dannhardt, G., Bauer, A., and Nowe, U.: Non-steroidal anti-inflammatory agents, Part 24. Pyrrolidino enaminones as models to mimic arachidonic acid, *Arch. Pharm. Weinheim.* **330** (1997), 74-82.

44 Boopathy, R. and Balasubramanian, A.S.: Purification and characterization of sheep platelet cyclo-oxygenase, *Biochem. J.* **239** (1986), 371–377.

45 Dragan, Y.P. and Ellis, E.F.: Optimization of an assay for studying the effects of agents on cyclooxygenase and lipoxygenase metabolism of arachidonic acid in washed human platelets, *Prostaglandins Leukot. Essent. Fatty Acids* **39** (1990), 105-109.

46 Johnson, J.L., Wimsatt, J., Buckel. S.D., Dyer, R.D., and Maddipati, K.R.: Purification and characterization of prostaglandin-H synthase from sheep placental cotyledons, *Arch. Biochem. Biophys.* **324** (1995), 26-34.

47 Brideau, C., Kargman, S., Liu, S., Dallob, A.L., Ehrich, E.W., Rodger, I.W., and Chan, C.C.: A human whole blood assay for clinical evaluation of biochemical efficacy of cyclooxygenase inhibitors, *Inflamm. Res.* **45** (1996), 68-74.

48 Cromlish, W.A. and Kennedy, B.P.: Selective inhibition of cyclooxygenase-1 and -2 using intact insect cell assays, *Biochem. Pharmacol.* **52** (1996), 1777-1785.

49 Ringbom, T., Noreen, Y., Perera, P., and Bohlin, L.: Development of a COX-1 and COX-2 in vitro assay for identification of natural products as inhibitors of prostaglandin biosynthesis, *J. Nat. Prod.* (1997) in press.

50 Wagner, H., Wierer, M., and Bauer, R.: In vitro-Hemmung der Prostaglandin-Biosynthese durch etherische Öle und phenolische Verbindungen, *Planta Med.* **52** (1986), 184-188.

51 Weithmann, K.U., Jeske, S., and Schlotte, V.: Effect of leflunomide on constitutive and inducible pathways of cellular eicosanoid generation, *Agents Actions* **41** (1994), 164-170.

52 Redl, K., Breu, W., Davis, B., and Bauer, R.: Antiinflammatory active polyacetylenes from *Bidens campylotheca, Planta Med.* **60** (1994), 58-62.

53 Pradelles, P., Grassi, J., and Maclouf, J.: Enzyme immunoassays of eicosanoids using acetylcholine esterase as label: an alternative to radioimmunoassay, *Anal. Chem.* **57** (1985), 1170-1173.

54 Maclouf, J., Grassi, J. and Pradelles, P.: Development of enzyme-immunoassay techniques for the measurement of eicosanoids, *Prostagl. Lipid Met. Rad. Inj.* (1987), 355-364.

55 Scioscia, K.A., Snyderman, C.H., Rueger, R., Reddy, J., D'Amico, F., Comsa, S., and Collins, B.: Role of arachidonic acid metabolites in tumor growth inhibition by nonsteroidal antiinflammatory drugs, *Am. J. Otolaryngol.* **18** (1997), 1-8.

56 Kuhl, P., Shiloh, R., Jha, H., Murawski, U., and Zilliken, F.: 6,7,4'-Trihydroxyisoflavan: a potent and selective inhibitor of 5-lipoxygenase in human and porcine peripheral blood leukocytes, *Prostaglandins* **28** (1984), 783-804.

57 Kuhl, P., Borbe, H.O., Fischer, H., Römer, A., and Safayhi, H.: Ebselen reduces the formation of LTB4 in human and porcine leukocytes by isomerisation to its 5S,12R-6-trans-isomer, *Prostaglandins* **31** (1986), 1029-1048.

58 Safayhi, H., Tiegs, G., and Wendel, A.: A novel biologically active seleno-organic compound - V. Inhibition by ebselen (PZ 51) of rat peritoneal neutrophil lipoxygenase, *Biochem. Pharmacol.* **34** (1985), 2691-2694.

59 Vasange-Tuominen, M., Perera-Ivarsson, P., Shen, J., Bohlin, L., and Rolfsen, W.: The fern *Polypodium decumanum* used in the treatment of psoriasis, and its fatty acid constituents as inhibitors of leukotriene B4 formation, *Prostaglandins Leukot. Essent. Fatty Acids* **50** (1994), 279-284.

60 Powell, W.S.: Reversed-phase high-pressure liquid chromatography of arachidonic acid metabolites formed by cyclooxygenase and lipoxygenases, *Analyt. Bioch.* **148** (1985), 59-69.

61 Veenstra, J., van de Pol, H., van der Torre, H., Schaafsma, G., and Ockhuizen, T.: Rapid and simple methods for the investigation of lipoxygenase pathways in human granulocytes, *J. Chromatogr.* **431** (1988), 413-417.

62 Wagner, H. and Fessler, B.: In vitro 5-lipoxygenase inhibition by *Elipta alba* extracts and the coumestan derivative wedelolactone, *Planta Med.* **52** (1986), 374-377.

63 Breton, J., Keller, P., Chabot-Fletcher, M., Hillegass, L., DeWolf, W.jr., and Griswold, D.: Use of a continuous assay of oxygen consumption to evaluate the pharmacology of 5-lipoxygenase inhibitors, *Prostaglandins Leukot. Essent. Fatty Acids* **49** (1993), 929-937.

64 Scioscia, K.A., Snyderman, C.H., Rueger, R., Reddy, J., D'Amico, F., Comsa, S., and Collins, B.: Role of arachidonic acid metabolites in tumor growth inhibition by nonsteroidal antiinflammatory drugs, *Am. J. Otolaryngol.* **18** (1997), 1-8.

65 Wagner, H.: Search for new plant constituents with potential antiphlogistic and antiallergic activity, *Planta Med.* **55** (1989), 235-241.

66 Alcaraz, M.J. and Ferrandiz, M.L.: Modification of arachidonic metabolism by flavonoids, *J. Ethnopharmacol.* **21** (1987), 209-229.

67 Ammon, H.P., Mack, T., Singh, G.B., and Safayhi, H.: Inhibition of leukotriene B4 formation in rat peritoneal neutrophils by an ethanolic extract of the gum resin exudate of Boswellia serrata, *Planta. Med.* **57** (1991), 203-207.

68 Safayhi, H., Mack, T., Sabieraj, J., Anazodo, M.I., Subramanian, L.R., and Ammon, H.P.: Boswellic acids: novel, specific, nonredox inhibitors of 5-lipoxygenase, *J. Pharmacol. Exp. Ther.* **261** (1992), 1143-1146.

69 Safayhi, H., Sailer, E.R., and Ammon, H.P.: Mechanism of 5-lipoxygenase inhibition by acetyl-11-keto-beta-boswellic acid, *Mol. Pharmacol.* **47** (1995), 1212-1216.

70 Sailer, E.R., Subramanian, L.R., Rall, B., Hoernlein, R.F., Ammon, H.P., and Safayhi, H.: Acetyl-11-keto-beta-boswellic acid (AKBA): structure requirements for binding and 5-lipoxygenase inhibitory activity, *Br. J. Pharmacol.* **117** (1996), 615-618.

71 Etzel, R.: Special extract of *Boswellia serrata* (H-15) in the treatment of rheumatoid arthritis, *Phytomedicine* **3** (1996), 91-94.

72 Gupta, I., Parihar, A., Malhotra, P., Singh, G.B., Ludtke, R., Safayhi, H., and Ammon, H.P.: Effects of Boswellia serrata gum resin in patients with ulcerative colitis, *Eur. J. Med. Res.* **2** (1997), 37-43.

73 Winking, M. and Simmet, Th.: Pharmacotherapeutic effects of extracts from *Boswellia serrata* in astrocytoma patients, Abstract of the 43rd Annual Congress on Medicinal Plant Research in Hall/Saale, 1995.

74 But, P.P.-H., Hu, S.-Y., and Kong, Y.C.: Vascular plants used in Chinese medicine, *Fitoterapia* **3** (1980), 245-264.

75 Pröbstle, A., Lotter, H., Wagner-Redecker, W., Matthiesen, U., and Bauer, R.: Cyclooxygenase inhibitory constituents from *Houttuynia cordata, Phytomedicine* **2** (1996), 305-308.

76 White, H.L. and Glassman, A.T.: Biochemical properties of the prostaglandin/thromboxane synthetase of human blood platelets and comparison with the synthetase of bovine seminal vesicles, *Prostaglandins* **12** (1976), 811-828.

77 Quoc, K.P. and Pascaud, M.: Effects of dietary gamma-linolenic acid on the tissue phospholipid fatty acid composition and the synthesis of eicosanoids in rats, *Ann. Nutr. Metab.* **40** (1996), 99-108.

78 Mirzoeva, O.K., Sud'ina, G.F., Pushkareva, M.A., and Varfolomeev, S.D.: Competitive inhibition of the 5-lipoxygenase-catalysed linoleate oxidation by arachidonic and 5-hydroperoxy-eicosatetraenoic acids, *FEBS Letters* **377** (1995), 306-308.

79 Samuelson, B.: Leukotrienes: mediators of immediate hypersensitivity reactions and inflammation, *Science* **220** (1983), 568-575.

80 Pröbstle, A., Neszmelyi, Jerkovich, G., Wagner, H., and Bauer, R.: Novel pyridine and 1,4-dihydropyridine alkaloids from *Houttuynia cordata, Natural Product Letters* **4** (1994), 235-240.

81 Nishikawa, Y.,Yasuda, I., Watanabe, Y., and Seto,T.: New polyacetylenic compounds in the rhizome of *Atractylodes lancea* DC. var. *chinensis, Yakugaku Zasshi* **96** (1976), 1322-1326.

82 Resch, M.S., Steigel, A., and Bauer, R.: Diacetoxy substituted polyacetylenes from *Atractylodes lancea, Phytochemistry* (1997), in press.

83 Pachaly, P., Lansing, A., and Sin, K.S.: Constituents of *Atractylis koreana, Planta Med.* **55** (1989), 59-61.

84 Resch, M.S., Steigel, A., Chen, Z.L., and Bauer, R.: 5-Lipoxygenase and cyclooxygenase-1 inhibitory active compounds from *Atractylodes lancea, J. Nat. Prod.* (1997), submitted.

85 Capon, J.R., Ghisalberti, E.L. and Jefferies, P.R.: Isoprenoid dihydroquinones from a brown alga, *Cystophora* sp., *Phytochemistry* **20** (1981), 2598-2600.

86 Heilmann, J., Resch, M., and Bauer R.: Antioxidant and radical scavenging activity of constituents from *Atractylodes lancea, Pharmaceutical Pharmacol. Lett.* (1997), submitted.

87 Morita, M., Nakajima, K., Taguchi, H., Yanagisawa, T., Sato, T., Chin, M., and Mitsuhashi, H.: New polyacetylenes from Bupleuri radix, Abstract of the 4th World Conference on Clinical Pharmacology and Therapeutics, Prague, 1989.

88 Alanko, J., Kurahashi, Y., Yoshimoto, T., Yamamoto, S., and Baba, K.: Panaxynol, a polyacetylene compound isolated from oriental medicines, inhibits mammalian lipoxygenases, *Biochem. Pharmacol.* **48** (1994), 1979-1981.

89 Bauer, R. and Wagner, H.: *Echinacea - Ein Handbuch für Ärzte, Apotheker und andere Naturwissenschaftler*, Wissenschaftliche Verlagsgesellschaft, Stuttgart, 1990.

90 Bauer, R., Remiger, P., and Wagner, H.: Alkamides from the roots of *Echinacea angustifolia, Phytochemistry* **28** (1989), 505-508.

91 Wagner, H., Breu, W., Willer, F., Wierer, M., Remiger, and Schwenker, G.: In vitro inhibition of arachidonate metabolism by some alkamides and alkylated phenols, *Planta Med.* **55** (1989), 566-567.

92 Müller, B., Redl, K., Breu, W., Pröbstle, A., Greger, H., and Bauer, R.: In vitro inhibition of cyclooxygenase and 5-lipoxygenase by alkamides from *Echinacea* and *Achillea* species, *Planta Med.* **60** (1994), 37-40.

93 Fish, F. and Waterman, P.G.: Rutaceae - Chloroform-soluble Alkaloids from the Root Bark of *Fagara chalybea, Phytochemistry* **11** (1972), 1866-1867.

94 Nakatani, M., Asai, H., Mochihara, K., and Hase, T.: Jatrorrhizine Chloride and other Constituents from *Fagara chalybea, Kagoshima Daigaku Rigakubu Kiyo, Sugaku, Butsurigaku, Kagaku* **23** (1990), 153, *ref. C.A.* **115**, 228433x..

95 Müller-Jakic, B.: PhD thesis, Munich, 1995.

96 Müller-Jakic, B., Pröbstle, A., and Bauer R.: In vitro inhibition of 5-lipoxgenase and cyclooxygenase by benzophenanthridine and protoberberine alkaloids, Poster at the 43rd Annual Congress on Medicinal Plant Research in Halle, 1995.

97 Dostal, J. and Potacek, M.: Quarternary benzophenanthridine alkaloids, *Collect. Czech. Res. Commun.* **55** (1990), 2840-2871.

98 Vavreckova, C., Gawlik, I. and Müller, K.: Benzophenanthridine alkaloids of *Chelidonium majus.* I. Inhibition of 5- and 12-lipoxygenase by a non-redox mechanism, *Planta Med.* **62** (1996), 397-401.

99 Wolff, J. and Knipling, L.: Antimicrotubule properties of benzophenanthridine alkaloids. *Biochemistry* **32** (1993), 13334-13339.

100 Vavreckova, C., Gawlik, I. and Müller, K.: Benzophenanthridine alkaloids of *Chelidonium majus*; II. Potent inhibitory action against the growth of human keratinocytes, *Planta Med.* **62** (1996), 491-494.

101 Deshpande, V.H. and Shastri, R.K.: Chemical investigation of three *Zanthoxylum* species, Z. *alatum*, Z. *oxyphyllum* and Z. *acconthopodium*, *Indian J. Chem.* **15**B (1977), 95-96.

102 Chen, J., Geng, G.L., Ye, G.H., Xi, F.D., Chen, S.F., He, C.H., and Zheng, Q.T.: Studies on the chemical constituents of *Zanthoxylum planispinum* Sieb. et Zucc., *Acta Pharm. Sinica* **23** (1988), 422-425.

103 Manandhar, M.D.: Isolation of pinoresinol dimethyl ether from *Zanthoxylum alatum* Roxb., *J. Nep. Chem. Soc.* **2** (1982), 19-21.

104 Müller-Jakic, B., Greger, H., Vermes, B., and Bauer, R.: Cyclooxygenase and 5-lipoxygenase inhibitory activity of tetrahydrofurofuran lignans, *Flavonoids and Bioflavonoids* (Eds. S. Antus, M. Gabor, K. Vetschera) pp. 149 - 156, Akademiai Kiado, Budapest, 1996.

105 Greger, H., and Hofer, O.: Unsymmetrically substituted tetrahydrofurouran lignans from *Artemisia absinthium* - Assignment of the Relative stereochemistry by lanthanide-induced chemical shifts, *Tetrahedron* **36** (1980), 3551-3558.

106 Brader, G.: Diploma thesis, University of Vienna, 1992.

107 Vermes, B., Seligmann, O., and Wagner, H.: Synthesis of biologically active tetrahydrofurofuranlignan-(Syingin, pinoresinol)-mono- and bisglucosides, *Phytochemistry* **30** (1991), 3087-3089.

108 Hutchinson, J.H., Riendeau, D., Brideau, C., Chan, C., Delirme, D., Denis, D., Falgueyret, J.-P., Fortin, R., Guay, J., Hamel, P., Jones, T.R., Macdonald, D., McFarlane, C.S., Piechuta, H., Scheigetz, J., Tagari, P., Therien, M., and Girard, Y.: Substituted thiopyrano[2,3,4-c,d]indoles as potent, selective, and orally active inhibitors of 5-lipoxygenase. Synthesis and biological evaluation of L-691,816, *J. Med. Chem.* **36** (1993), 2771-2787.

109 Umeda-Sawada, R., Takahashi, N., and Igarashi, O.: Interaction of sesamin and eicosapentaenoic acid against delta 5 desaturation and n-6/n-3 ratio of essential fatty acids in rat, *Biosci. Biotechnol. Biochem.* **59** (1995), 2268-2273.

110 Hong, M.F., Pan, Y.X., Jiang, T.Y., Li, C.L., and Han, G.Q.: Isolation and characterization of platelet activating factor (PAF) inhibitors from *Zanthoxylum schinifolium*, *J. Chin. Pharm. Sci.* **1** (1992), 13-19.

111 Hong, M.F., Pan, J.X., Hao, M., Yao, B., and Han, G.: The active constituents of platelet activating factor (PAF) inhibitors from *Zanthoxylum planispinum* Sieb. & Zucc., *Zhiwu Ziyuan Yu Huanjing*, **2** (1993), 25-7; *ref. C.A.* **119**, 85668x.

112 Ferrandiz, M.L., Nair, A.G.R., and Alcaraz, M.J.: Inhibition of sheep platelet arachidonate metabolism by flavonoids from spanish and indian medicinal herbs, *Pharmazie* **45** (1990), 206-208.

113 Lyckander, I.M. and Malterud, K.E.: Lipoxygenase Inhibiting Flavonoids from *Orthosiphon spicatus*, in Das, N.P. (ed.) *Flavonoids in Biology & Medicine III: Current Issues in Flavonoid Research*, pp. 469-474, National University of Singapore, 1990.

114 Ferrandiz, M.L. and Alcaraz, M.J.:Anti-inflammatory activity and inhibition of arachidonic acid metabolism by flavonoids, *Agents Actions* **32** (1991), 283-288.

115 Berenbaum, M.C.: Isobolographic, algebraic, and search methods in analysis of multiagent synergy, *J. Am. Coll. Tox.* **7** (1988), 927-937.

116 Sühnel, J.: Evaluation of synergism or antagonism for the combined action of antiviral agents, *Antiviral Res.* **13** (1990), 23-40.

117 Liu, J.H., Xu, S.X., Yao, X.S., and Kobayashi, H.: Two new 6-alkylcoumarins from *Angelica pubescens f. biserrata*, *Planta Med.* **61** (1995), 482-484.

118 Liu, J.H., Xu, S.X., Yao, X.S., and Kobayashi, H.: Angelol-type comarins from *Angelica pubescens f. biserrata* and their inhibitory effect on platelet aggregation, *Phytochemistry* **39** (1995), 1099-1101.

119 Liu, J.H., Xu, S.X., Yao, X.S., and Kobayashi, H.: Further isolation of chemical constituents from *Angelica pubescens f. biserrata*, *Acta Pharm. Sinica* **31** (1996), 63-67.

120 Liu, J.H., Zschocke, S., and Bauer, R.: A polyacetylenic acetate and a coumarin from *Angelica pubescens f. biserrata*, *Phytochemistry* (1997), accepted.

121 Liu, J.H., Zschocke, S., Reininger, E., and Bauer, R.: Inhibitory effect of *Angelica pubescens f. biserrata* on 5-lipoxygenase and cyclooxygenase, *Planta Med.* (1997), submitted.

122 Yang, X.-W., Gu, Z.-M., Wang, B.-X., Hattori, M., and Namba, T.: Comparison of anti-lipid peroxidative effects of the underground parts of *Notopterygium incisum* and *N. forbesii* in mice, *Planta Med.* **57** (1991), 399 - 402.

123 Zschocke, S., Seibt, A., Stöhr, J., and Bauer, R.: The influence of fatty acids and synergistic effects on the 5-lipoxygenase inhibitory activity of Qianghou (*Notopterygium incisum*), Poster at the 44th Annual Congress of the Society for medicinal Plant Research in Prague, 1996.

124 Zschocke, S.: PhD thesis University of Düsseldorf, in preparation.

125 Sud'ina, G.F., Mirzoeva, O.K., Pushkareva, M.A., Korshunova, G.A., Sumbatyan, N.V., and Varfolomeev,

S.D.: Caffeic acid phenethyl ester as a lipoxygenase inhibitor with antioxidant properties, *FEBS Lett.* **329** (1993), 21-24.

126 Natarajan, K., Singh, S., Burke, T.R.jr., Grunberger, D., and Aggarwal, B.B.: Caffeic acid phenethyl ester is a potent and specific inhibitor of activation of nuclear transcription factor NF-kappa B, *Proc. Natl. Acad. Sci. USA* **93** (1996), 9090-9095.

127 Yamamoto, K., Arakawa, T., Ueda, N., and Yamamoto, S.: Transcriptional roles of nuclear factor kappa B and nuclear factor-interleukin-6 in the tumor necrosis factor alpha-dependent induction of cyclooxygenase-2 in MC3T3-E1 cells, *J. Biol. Chem.* **270** (1995), 31315-31320.

128 Crofford, L.J., Tan, B., McCarthy, C.J., and Hla, T.: Involvement of nuclear factor kappa B in the regulation of cyclooxygenase-2 expression by interleukin-1 in rheumatoid synoviocytes, *Arthritis Rheum.* **40** (1997), 226-236.

129 Chen, F., Sun, S., Kuhn, D.C., Gaydos. L.J., Shi, X., Lu, Y., and Demers, L.M.: Involvement of NF-kappa B in silica-induced cyclooxygenase II gene expression in rat alveolar macrophages, *Am. J. Physiol.* **272** (1997), L779-786.

130 Lyss, G., Schmidt, J.T., Merfort, I., and Pahl, H.L.: Helenalin, an anti-inflammatory sesquiterpene lactone from Arnica, selectively inhibits transcription factor NF- B, *Biol. Chem.* **378** (1997), 951-961.

131 Hwang, D., Fischer, N.H., Jang, B.C., Tak, H., Kim, J.K. and Lee, W.: Inhibition of the expression of inducible cyclooxygenase and proinflammatory cytokines by sesquiterpene lactones in macrophages correlates with the inhibition of MAP kinases, *Biochem. Biophys. Res. Com.* **226** (1996), 810-818.

132 Tordjman, C., Coge, F., Andre, N., Rique, H., Spedding, M., and Bonnet, J.: Characterisation of cyclooxygenase 1 and 2 expression in mouse resident peritoneal macrophages in vitro; interactions of non steroidal anti-inflammatory drugs with COX-2, *Biochim. Biophys. Acta* **1256** (1995), 249-256.

133 Creminon, C., Habib, A., Maclouf, J., Pradelles, P., Grassi, J., and Frobert, Y.: Differential measurement of constitutive (COX-1) and inducible (COX-2) cyclooxygenase expression in human umbilical vein endothelial cells using specific immunometric enzyme immunoassays, *Biochim. Biophys. Acta* **1254** (1995), 341-348.

134 Zschocke, S., Lehner, M., Bauer, R.: 5-Lipoxygenase and cyclooxygenase inhibitory constituents from Qianghuo (*Notopterygium incisum*), *Planta Med.* **63** (1997), 203-206.

NATURAL PRODUCTS IN DRUG DISCOVERY AND DEVELOPMENT

W.N.A. ROLFSEN SANDSBORG
Division of Pharmacognosy, Department of Pharmacy, Biomedical Center, Uppsala University, Box 579, S-751 23 Uppsala, Sweden and Phase I Services, Quintiles AB, Islandsgatan 2, S-753 18 Uppsala, Sweden

Abstract

The revolution in biology over the past two decades has resulted in radical new opportunities for discovery of new drugs. Most importantly, it has defined major drug targets in the form of molecular components of disease processes.

In recent years, there is a growing pressure on the pharmaceutical industries to increase the efficiency with which novel leads are generated and screened for biological activity.

Natural products have served as a major source of drugs for centuries, and about half of the pharmaceuticals in use today are derived from natural products.

Nature is unrivalled in its ability to produce organic molecules with structural complexity and biological potency; thus nature has been performing combinatorial chemistry for thousands of years.

Interest in natural products research within the new era of drug research is strong and can be attributed to several factors. These include issues like unmet medical needs, diversity of both chemical structures and biological activities, the utility of bioactive natural products as biochemical and molecular probes, the development of novel and sensitive techniques to detect biologically active natural products, improved techniques to isolate, purify, and structurally characterise the active constituents.

The opportunities within the new era of drug discovery and development, for multidisciplinary research that joins the forces of natural product chemistry, molecular and cellular biology, synthetic and analytical chemistry, biochemistry, and pharmacology to exploit the diversity of chemical structures and biological of natural products, are immense.

Introduction

Pressure on the pharmaceutical industry has steadily grown to cut research and development costs, while increasing the speed of bringing in new drugs to market.

Thus, there is a need for increasing the efficiency with which novel leads are generated and screened for biological activity.

The goal for many research groups in most pharmaceutical industries are to produce candidate drugs more quickly and to ensure that the candidates survive development.

New approaches to treat are sought intensively by research organisations within the pharmaceutical industry and elsewhere.

Redesigning drug discovery

In order to bring new drugs to the market faster, the traditional drug discovery and development is being reorganised (Fig. 1).

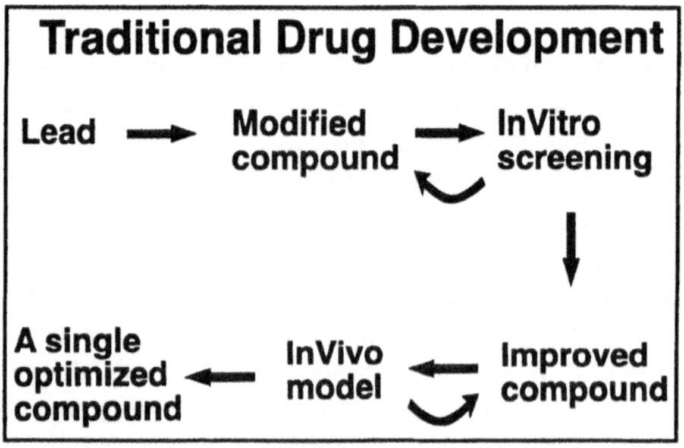

Fig.1. Traditional drug discovery and development.

Organisations continue to seek to improve the strategies by which we discover new drugs (1). Drug discovery demands a co-ordinated effort among people with different talents. Elucidation of disease mechanisms and looking for good targets walk hand in hand with those skilled in chemistry or isolation of new molecules and those who can determine safety and effectiveness of the molecules.

The pressure on the pharmaceutical industry to produce more effective candidate drugs has initiated a *redesign of drug discovery* of today.

A key step in the drug discovery process is the identification of a disease target and a demonstration that its activity affects the disease process. Although molecular biology has played a role in drug discovery earlier, the wide availability of data bases for gene and protein sequences has raised from an rare speciality to a necessary tool. With this change an increasing requirement for bioinformatics has come, if researchers are to manipulate and compare sequence data in the search for similarities, targets and models.

This validation of gene targets demands sophisticated computing if sequence and structure data are to be integrated, molecules are to be modelled and patient data are to be accessible.

The expressed protein needs to be identified and also maybe biotechnically produced in order to obtain optimal models for screening. When the target is identified and isolated/produced, the screening system is set up.

High Troughput Screening

Five years ago, High Throughput Screening was largely a manual process. Today, researchers can test 30 times as many samples for basic biological activity, largely due to investments in robotics and miniatinisation.

High Throughput Screening (HTS) is today the process by which a large number of compounds can be tested in an automated fashion, for inhibition or activation of a particular biological target (2).

The primary goal is to identify high quality leads, active in a fairly low concentration and if possible with a new type of structure.

The most important role of HTS is to detect leads and supply direction for their optimisation. A well designed HTS can provide information in addition to the potency of a compound. Information on specificity can be obtained by running concomitantly a counter-screen with a related target.

HTS cannot evaluate bioavailability, *in vivo* pharmacology and toxicity. Thus, medicinal chemistry and pharmacology studies are required to convert a compound that emerges from HTS to a useful drug. The greater the number and diversity of compounds that are run through the screen, the more successful it is likely to be.

Combinatorial Chemistry

In order to create a great number and diversity of compounds, many companies use the possibility of combinatorial chemistry. Directed combinatorial chemistry relies on producing all possible combinations of a basic set of modular components. Large libraries consisting of all possible combinations of substances are often the basis of the screening (3).

Structure-based drug design

To define the topographies of the complementary surfaces of ligands and their macromolecular targets is another important part of drug discovery. The three dimensional structures of more than 4000 macromolecules have already been solved, and the number will continue to increase steadily. Many of these macromolecules are important drug targets and it is now possible to use the knowledge of their three dimensional structure as a good base for drug design (4).

Balance in drug discovery

Drug discovery teams earlier produced compounds with potency and selectivity for particular receptor classes only to find them dropped when pharmacokinetics and toxicity were assessed. Its now accepted that one may have to compromise potency in order to achieve a balance between potency, pharmacokinetics and safety and that these areas needs to come in as early as possible in the discovery process.

Of course one has to bear in mind that due to the new ways of designing new drugs, many of the new techniques are not yet fully tested. It is important to remember that there are parts of the traditional drug discovery techniques that can still produce new and interesting compounds.

What then is the role of natural products in this brave new drug discovery world?

Natural products have served as a major source of drugs for centuries, and about half of the pharmaceuticals in use today are derived from natural products (5).

Nature is unrivalled in its ability to produce organic molecules with structural complexity and biological potency; thus nature has been performing combinatorial chemistry for thousands of years.

What is then the interest of natural products in modern drug discovery? Are there any new drugs developed lately on the basis of pharmacognosy/natural products.

Over the past ten years, there has been a resurgence of interest in the investigation of natural materials as a source potential. There are now signs that this interest is once more in favour of new approaches to drug discovery. In order to provide a more solid basis for the claims made for the importance of natural products in drug discovery and development, data of new drugs approved by either the United States Food and Drug Administration (FDA) or comparable entities in other countries, have been analysed. Cragg and coworkers (6) analysed the number and sources of approved drugs, reported mainly in the Annual Reports of Medicinal Chemistry for the years 1983-1994, predominantly in the area of cancer and infectious diseases. Other data from a variety of sources was also covering potential anticancer compounds reported to be in the preNew Drug Application (NDA) phase up to the end of 1995.

Drugs of natural origin have been classified as original natural products, products derived semisynthetically from natural products, or synthetic products based on natural product models.

The data has been analysed in terms of numbers classified according to their origin as indicated below:

B: Biologics (vaccines, monoclonals, etc derived from mammalian sources).
N: Derived from unmodified natural product source.
ND: Derived from a natural product source (e.g., semisynthetic).
S: Exclusively from a synthetic source.
S*: From a synthetic source, but originally modelled on a natural product parent.

The data are presented in the form of bar graphs and tables for the following categories;

- New Approved drugs: 1983-1994 (Fig. 2).
- Available Anticancer Drugs Through 1994 (Fig. 3).
- Pre NDA Anticancer Drugs 1989-1995 (Fig. 4).

Of the new approved drugs reported between 1983 and 1994 (Fig. 2) drugs of natural origin (N and ND) predominate (78%) in the areas of antibacterials, while 61% of the 31 anticancer drugs are naturally derived (N and ND) or modelled on a natural product parent (S*).

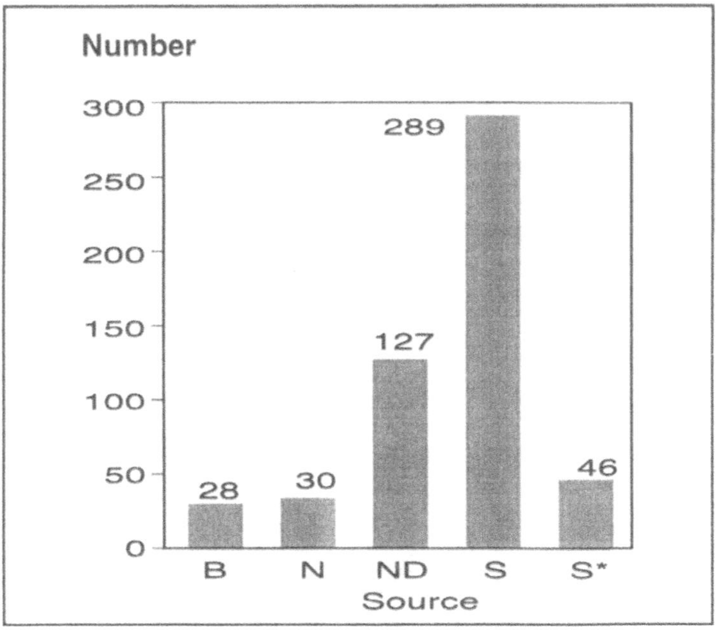

Fig. 2. New approved drugs: 1983 - 1994.

Of the 87 approved anticancer drugs (Fig. 3), 62% are of natural origin or modelled on natural product parents.

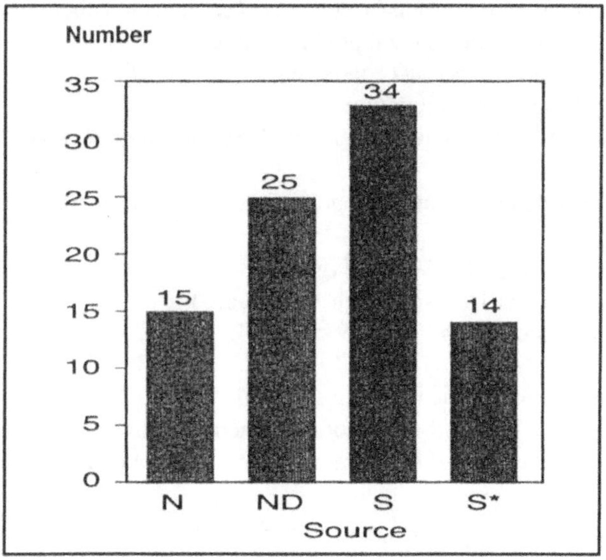

Fig.3. Available anti-cancer drugs: through 1994.

Of the 299 pre-NDA anticancer drug candidates (Fig.4) (i.e. in pre-clinical development or clinical development) for the period 1989-1995, 50 are original natural products and 48 semisynthetic derivators, while 30 are based on natural product models. The percentage of the drugs having natural origin, excluding the biologics, is 61 %,

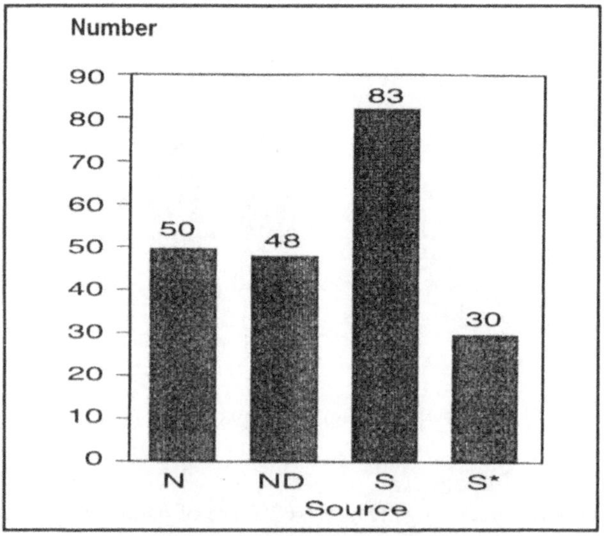

Fig. 4 Pre NDA Anticancer Drugs 1989 - 1995.

Cragg and co-workers high-lights the invaluable role that the natural products have played, and continue to play, in the drug discovery and development of the future. The importance of natural products in the future of drug discovery is clear: novel biological active natural products will continue to serve as lead compounds for drug development and as biochemical probes for the discovery of pharmacological and biochemical processes.

Dramatic advances in cell culture and extraction techniques, analytical techniques, powerful techniques in structure elucidation, HTS, genetics and protein biochemistry, structural biology and synthetic chemistry have converted to create a bright future at the crossroads of natural product science and drug discovery.

Conclusion

To summarise, in the pharmaceutical industry there is a need for innovative new drug candidates with novel structures and biological action and with the possibility of opening up new areas of therapy. We stand on the threshold of a new area in pharmaceutical industry. Target identification will come primarily from genomics and basic cell biology research, aided by new chemicals from natural products or classical chemistry that define new pathways.

The advancement in new techniques in the drug discovery process, together with the strong pressure of bringing in new chemical leads for drug development will position natural products as an important source of lead compounds in drug discovery programs.

References

1) Caporale, L.H. (1995) A view from the pharmaceutical industry. Chemical ecology, *Proc.Natl.Sci. USA* **92**, 75-82.
2) Broach, J.R. and Thomer, (1996) High Throughput Screening for drug discovery, *Nature* **384** Suppl.
3) Hogen, C.J. (1996) Directed combinational chemistry, *Nature* **384** Suppl.
4) Blundell, T.M. (1996) Structure based drug design, *Nature* **384** Suppl.
5) Clark, A.M. (1996) Natural products as a resource for new drugs, *Pharmaceutical research* **13** (No. 8)
6) Cragg, G.M. Newman, D.J. and Snader, K.M. (1997) Natural Products in drug discovery and development *J. Nat. Prod.* 60(52).

QPRI'S SYSTEM FOR SCREENING OF NATURAL PRODUCTS.

R. J. QUINN
Queensland Pharmaceutical Research Institute, Griffith University, Brisbane, 4111, Australia

The Queensland Pharmaceutical Research Institute (QPRI) has been undertaking a natural product drug discovery project with Astra. Screening commenced on 1 July 1994.

The natural product drug discovery project at QPRI has been examining the biodiversity of Queensland. The plant collection programme is undertaken by the Queensland Herbarium who have extensive expertise in the area and an outstanding herbarium collection. There are around 9000 plants in Queensland of which 75% are endemic to Australia.The marine collection programme is undertaken by the Queensland Museum who have a particular expertise in sponge taxonomy. There is an estimate of 5,000 macro benthic marine invertebrates in the waters off the Queensland coast. The biological diversity translates to chemical wealth as plants and marine organisms are chemical factories and the diversity of the species results in a diversity of compounds. This is an untapped resource for new lead discovery.

The natural product drug discovery programme in our laboratory consists of plant and marine collection, extraction, high throughput screening of the extracts, identification

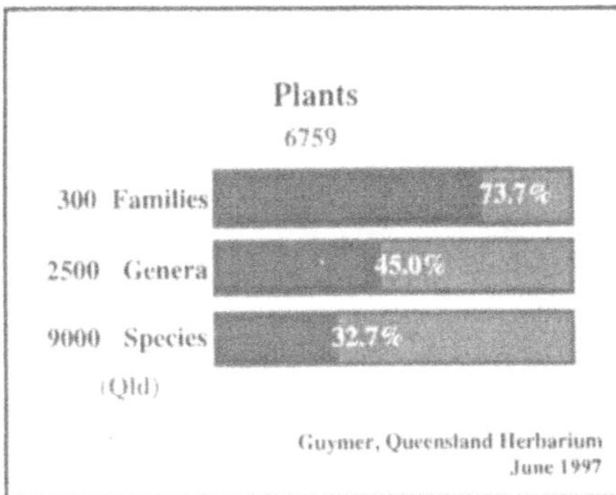

Figure 1 Plants collected to date showing percentage of Queensland flora

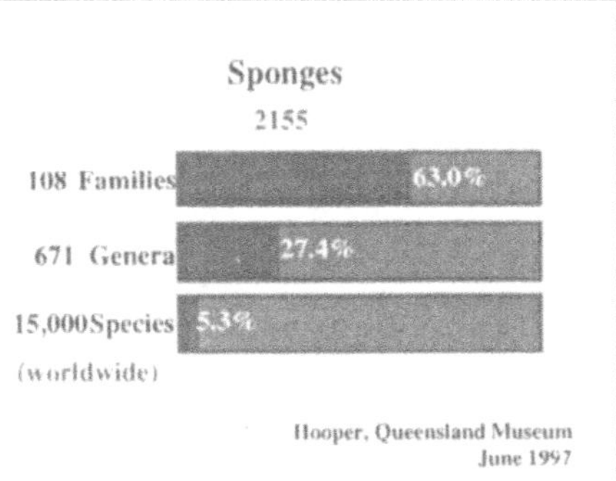

Figure 2 Marine sponges collected to date showing percentage of estimated world sponge population

of active extracts, bioassay directed fractionation, isolation of the active constituents and structure elucidation.

QPRI has established a diverse natural product collection of plants and marine organisms, with 9198 and 2623 samples, respectively, as at June 1997. The 9198 plant samples cover 221 of the 300 families, 1125 of the 2500 genera and 33% or 2946 of the 9,000 species in Queensland. The majority of the marine collection (2155 samples of the 2623) are sponges and cover 68 of the known 108 families worldwide, 184 of the 673 genera but only 5.3% or 790 of the estimated worldwide 15,000 species.

In order to be successful in any screening operation the vast amount of data generated by HTS of extracts, fractions, or compounds must be harnessed. HTS has a high continuous flow of information from many sources, such as sample data, extract data and screening data. Screens may be run in parallel so that assay results from various measuring instruments may be obtained on the same sample. This means there needs to be multiple contact points with data and multiple treatments of data. The data must be available at all times. The software must allow the flexibility to change data flow as future options are not always known or predicted. The data must be able to be queried and the data analysed in various ways. HTS requires a data base running on a fast server with multiple-user access to the same data. The database must be able to be updated simultaneously. There must be robust backup and maintenance of the server. Once the data base can capture and store the data, it is only as good as the query tools that allow access to the data. We have opted to write our own software. This decision was based on evaluation of commercial packages and a decision that in-house development would allow as good or better customisation in the initial stages but, critically, would allow easier adaptation to future unforeseen requirements.

The software package, QPRI*HiTbaSe*, has served us extremely well. QPRI*HiTbaSe* is a client / server based system with graphical client software for data entry and query. All information is stored in SQL tables on an Oracle server. Assays can be set-up graphically on a microtitre plate format with mouse-click control. Raw screening data from instrumen-

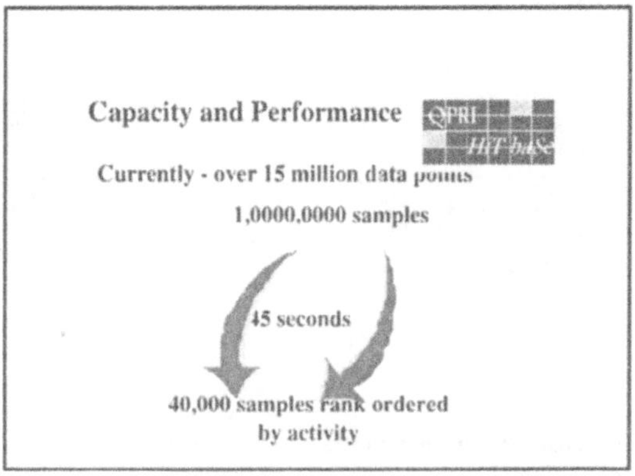

Figure 3 The search performance of QPRI*HiTbaSe*

tal readout can be downloaded from the instruments' computers to the database in batches of up to 512 plates (49,1521 individual results) from an unlimited number of individual computers at a rate of 5 plates (480 results) per minute. The software has in-built statistical treatment for quality control of assays. Raw data is converted to an activity index and hit selection performed automatically using pre-selected criteria. All data can be re-analysed at any time using different criteria for hit selection. The software can provide working lists of active extracts and tracking of follow-up assays. There is complete management of working lists. In regards to capacity and performance QPRI*HiTbaSe* has over 15 million data points and a query of 1 million samples to produce a list, rank ordered by activity, of 40,000 samples took 45 seconds.

High Throughput screening at QPRI has focused on receptors, enzymes and mechanism based cellular assays in the area of cardiovascular, respiratory, gastrointestinal, autoimmune diseases, pain control, bacterial and parasite infection, neurodegenerative and mental disease.

The natural product drug discovery project uses a bioassay guided fractionation to isolate active compounds from identified active extracts. At the commencement of the project we envisaged the need to isolate only 1 mg of material for structure elucidation and biological evaluation in the screen of interest.

The requirement for a small amount of isolated compound established the logistics of our natural product drug discovery project. Small scale extraction and chromatography were integral to the approach. The gains in efficiency with miniaturization have been substantial, allowing the isolation of over 200 bioactive compounds in the past three years. The structures of over 50% of compounds have been obtained within 24 hours using COSY, gHMQC and gHMBC on a 600 MHz NMR spectrometer. More involved NMR experiments have given the structures of all the remaining compounds.

At the commencement of the project we evaluated 10,000 extracts per year in up 35 different screens per year. In 1995-96, the number of assays performed on extracts was 534,000. Hit rates have varied with the screen and is dependent on concentration of extracts and criteria for activity, In one example, 21,162 extracts produced 9 hits (active extracts) or a hit rate of 0.04%. In this case, we isolated 12 active compounds from the 9 active extracts that were fractionated. In 1994-95, we isolated 34 bioactive compounds, in 1995-96 we isolated 61 bioactive compounds and in 1996-97 we isolated 110 bioactive compounds. In the first 205 compounds isolated, there were 50 classes of compounds and 71 new compounds, including 5 new structural classes. The natural products have thus displayed diversity in structural type. The known compounds have demonstrated biological properties not previously reported.

The process has identified many potential lead compounds, however this has required a screen lifetime of two years. The evaluation of 20,000 extracts took two years and this is not in pace with the need for rapid identification of leads. The most interesting or potent extract may not be identified until the later part of the screening.

During the first two years, QPRI was acummulating an extract library. This extract library now has over 30,000 extracts.

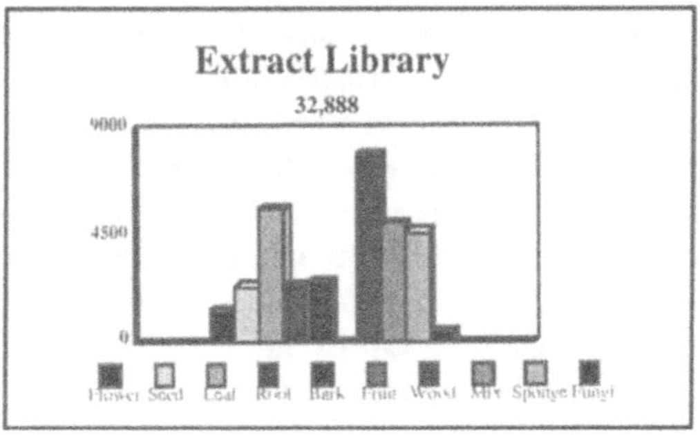

Figure 4 The QPRI extract library showing the number of extracts of various plant parts, marine organisms and fungi

The QPRI strategy is now to evaluate the extract library and identify the bioactive constitiuents in about a six month period. One such project commenced on 5 June 1996 and finished on 28 March 1997 and resulted in the identification of 30 enzyme inhibitors with IC_{50}s of 10 μM or less. Another project that commenced on 17 December 1996 and finished on 30 June 1997 resulted in the identification of 3 compounds with better than IC_{50} 3 μM inhibition.

I would now like to describe the isolation and structure elucidation of the first compound to come from the natural product high throughput screening project.

The methanol extract of flowers of a *Elaeocarpus* species displayed activity of 89% and 98% for inhibition of lymphocyte proliferation activated respectively by SEA and IL-2. Bioassay guided fractionation of the methanol extract using C-18 reverse phase HPLC resulted in the isolation of an active compound elaeocarpin-A. Larger scale work-up provided elaeocarpin-A (0.75 % dry weight) and the related compound elaeocarpin-B (0.44 % dry weight).

Elaeocarpin-A had a molecular weight of 690 (positive mode electrospray mass spectrum) consistent with a molecular formula of $C_{35}H_{26}O_{16}$. The 600 MHz NMR HMBC experiment provided enough information to solve the structure. A correlation from a methylene proton of a sugar at d 3.92 to the anomeric carbon at 95 ppm established that the 5'-methylene was in a ring and that the sugar was a pyranoside. The anomeric proton at d 5.72 showed a correlation to a carbon at 163 ppm. This carbon also correlated to two *meta* coupled aromatic protons at d 6.25 and d 6.32, establishing the point of attachment of the sugar to a flavone. An identified hexahydroxyphenoyl group was bound to the sugar via ester linkages at C-2' and C-3' of the sugar as HMBC correlations were observed between d 4.96 and the carbonyl carbon at 167 ppm and d 4.78 to the carbonyl carbon at 168 ppm.

Elaeocarpin-B also had a molecular weight of 690. Elaeocarpin-B was related to elaeocarpin-A however the methylene protons at d 2.85 and a methine proton at d 5.68 in elaeocarpin-A became doublets (J = 16.2 Hz) at d 8.01 and d 7.69 establishing a *trans* double bond. The meta coupled protons at d 6.28 and d 6.20 in elaeocarpin-A became a two proton singlet at d 6.11 establishing that symmetry had been introduced.

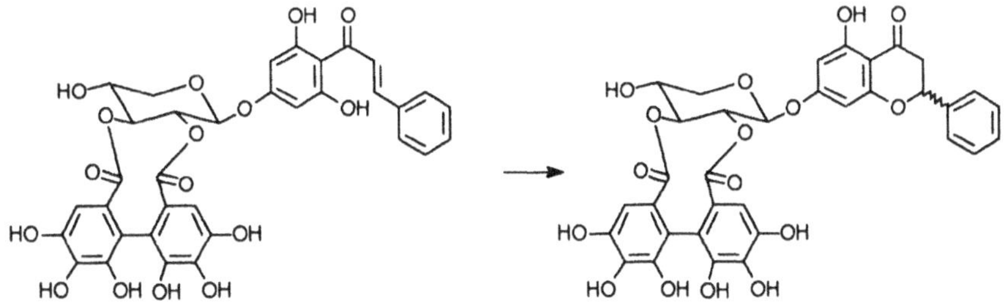

elaeocarpin-B elaeocarpin-A

Detailed analysis of the ^{1}H NMR spectrum of elaeocarpin-A indicated that six protons were doubled (d 2.86, 2.85; 6.26, 6.25; 6.32, 6.31; 5.73, 5.72; 3.63, 3.62; 3.920, 3.923). Elaeocarpin-B cyclised in either methanol or dimethyl sulphoxide to elaeocarpin-A. NMR analysis of the crude methanol extract indicated that elaeocarpin-A was present in only small amount compared to elaeocarpin-B. Elaeocarpin-A appears to be a 1:1 mixture of epimers at C2 in the flavone.

Both elaeocarpin-A and -B had similar activity (IC_{50} 2.4 - 12.2 mM) in inhibiting lymphocyte proliferation following activation by either SEA or IL-2.

	% dry weight	**lymphocyte proliferation IC_{50} (mM)**	
		SEA	**IL-2**
MeOH extract - 100 mg / well		89%	98%
elaeocarpin-A	0.75	3.3	3.9
elaeocarpin-B	0.44	2.4	12.2

Human plasminogen activator inhibitor type-1 (PAI-1), a member of the serine protease inhibitor family is the main inhibitor of human tissue plasminogen activator (tPA) in plasma. The cleavage of plasminogen by tPA governs the release of the active enzyme plasmin, which degrades the fibrin clot. PAI-1 binds irreversibly to tPA and totally inhibits tPA activity, and thereby all fibrinolytic activity.[1,2] Clinical studies have shown that high levels of plasma PAI-1 may be related to the pathogenesis of arterial and thrombotic diseases.[2,3] Specific inhibition of PAI-1 should therefore be a novel way to enhance fibrinolysis.

A total of 21,384 extracts of plant and marine organisms were screened for inhibition of PAI-1 in the presence of tPA. The dichloromethane extract of the flowers of *Eucalyptus albens* (Myrtaceae) was active at 500 mg/mL (dry weight). Bioassay guided fractionation led to the isolation of Sideroxylonal C (**1**), a previously undescribed isomer of Sideroxylonal A and B.[4]

Flash chromatography of the dichloromethane extract (1.417 g) on silica (stepped gradient elution from 100% hexane to 100% dichloromethane, and then 100% ethyl acetate) yielded ten fractions. Active fractions 7 and 8 were combined (280.9 mg) and chromatographed on Sephadex LH-20 with methanol as eluant giving 5 subfractions. Subfraction 4 contained Sideroxylonal C (9.7 mg, 0.06%).

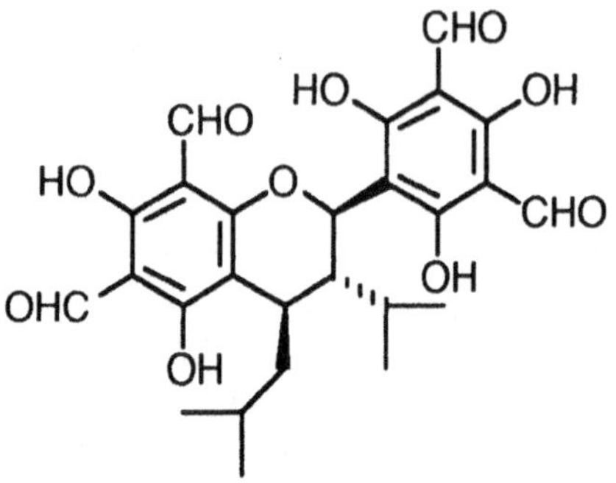

sideroxylonal C

The relative stereochemistry was determined by analysis of NOESY experiments and the coupling constants around the three chiral centres C7', C10' and C7, and considerations of molecular models.[5] The coupling constant $J_{H7'-H10'}$ = 10.8 Hz indicated that H7' and H10' were trans-diaxial. The coupling constant $J_{H10'-H7}$ was 2.4 Hz. Molecular modelling predicted that the lowest energy conformation was one in which the six-membered ring heterocycle in sideroxylonal C assumed a distorted boat with calculated angles of 154° between H7' and H10' and 112° between H10' and H7. A coupling constant of 2.4 Hz was also observed between H10' and H11' indicating restricted rotation around the C10'-C11' bond in the lowest energy conformation which had a dihedral angle of 70° between H10' and H11'.

Further confirmation for the relative stereochemistry was obtained from NOESY experiments. Strong NOE enhancements were observed from H7 to H10a, H10b, H12' and H13', and from H7' to H11', H12' and H13' which is consistent with the proposed relative stereochemistry.

Sideroxylonal C was tested against PAI-1 in the presence of Flavigen™ and tPA and showed an IC_{50} = 5 μM.

Acknowledgments

I would like to thank the natural product drug discovery team Tony Carroll, Roger Moni, Priscila Leone, Mark Butler, Greg Pierens, Sylvia Urban, Shaun Tennant, Greg Fechner,

Joanne Redburn, Peter Walve, Lekha Suraweera, Anna Ngo, Rama Addepalli, Brett, Lisa Harris, Jill Smith, Jasmine Lamb, Caroline Tranter, Rolf Hopkins, Sue Kydd, Fiona Stewart, Anthony Boyle, Earl Douglas, Tina Walsh, Marni Burkhardt, Kimberley Myers, Gordon Clement, Lynda Macfarlane, Dianne Hargans, Christina Lien, Lesley Layton.

I would like to acknowledge the collaboration of of Gordon Guymer, Paul Forster, Tim Ryan and Laurie Jessup from the Queensland Herbarium for collection and identification of plants and John Hooper, Steven Cook, John Kennedy, Paula Tomkins, Sue List and Elizabeth O'Brien form the Queensland Museum for collection and identification of marine organisms.

Specifically, the isolation and structure elucidation work on elaeocarpin-A and -B was performed by Tony Carroll and of sideroxylonal C by Priscila Leone.

The support of Astra Pharmaceuticals Pty Ltd. is gratefully acknowledged

References and Notes

1. Krishnamurti, C.; Alving, B. M. *Seminars in Thrombosis and Hemostasis*, **1992**, *18*, 67.
2. Verstraete, M.; Collen, D. *Blood*, **1986**, *67*, 1529.
3. Dawson, S.; Henney, A. *Atherosclerosis* **1992**, *95*, 105.
4. Satoh, H.; Etoh, H.; Watanabe, N.; Kawagishi, H.; Arai, K.; Ina, K. *Chem. Letters.*, **1992**, 1917 1920.
5. Molecular modelling was carried out using SYBYL on a silicon graphics computer with energy minimisations using the Powell method.

PHYTERA'S STRATEGY FOR THE DISCOVERY OF NOVEL ANTI-INFECTIVE AGENTS FROM PLANT CELL CULTURES

JAMES B. MCALPINE*, CHRISTOPHER PAZOLES*
AND ANGELA STAFFORD**
**Phytera, Inc., 377 Plantation Street, Worcester, MA 01606 USA*
***Phytera Ltd., Regent Court, Regent Street, Sheffield, S14DA, UK*

Phytera is a young biopharmaceutical company with facilities in Copenhagen, Denmark; Sheffield, UK and Worcester, Massachusetts and with the mission of building the World's premier natural products based drug discovery company. Our rationale is based on the role played by natural products as drugs or as lead molecules from which developed structure activity relationships has lead to a new drug. Cragg, et al[1] in an analysis of new chemical entities entering the US approved drug market in the period 1983-1994 pointed out that 249 were either natural products or were derivatives or analogs of natural products. By contrast 286 new chemical entities entering the market in that decade owed their origin to chemical library screening or designed synthesis (Figure 1). This slight preponderance of synthetic origin must be viewed in the relative efforts in pharmaceutical research between medicinal and natural product chemistry. Most large pharmaceutical companies have at least ten fold more research effort devoted to medicinal chemistry then natural product discovery. If we look at the area of anti-infective therapy the absolute numbers favor the natural products. Here 59 new products for the decade have natural product parentage compared with 34 with medicinal chemistry.

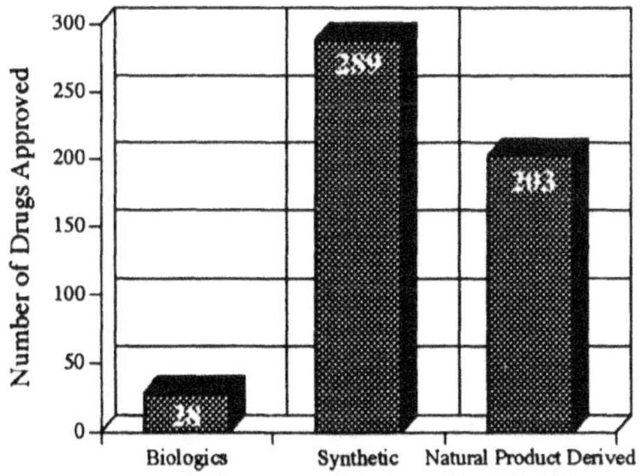

Figure 1. New USFDA Approved Drugs for 1983-1994 - Source of Pharmacophore

Phytera has chosen within the natural products area two fields of concentration. These both represent under exploited endeavors with the further advantage of assured reaccess; marine derived microbial cultures and plant cell cultures. This article will describe in broad detail our strategy for drug discovery from plant cell culture.

Although the basic complexity of plant species vary widely, higher plants are among the most genomically rich organisms on the planet. A typical bacterium contains about 1,000 genes while a fungus has about an order of magnitude more and most mammals about 100,000. The cells of higher plants can contain as many as 250,000 genes. Given that most secondary metabolites represent the second generation products of several genes one might expect that sequential increase in the number and complexity of secondary metabolites might parallel a simple arithmetic increase in the number of genes. Yet this does not seem to be the case. One reason for the relative paucity from plants is demonstrated by a survey of taxol content in samples of *Taxus brevifolia* collected in different locations in the U.S. Pacific Northwest[2]. The results are shown in figure 2.

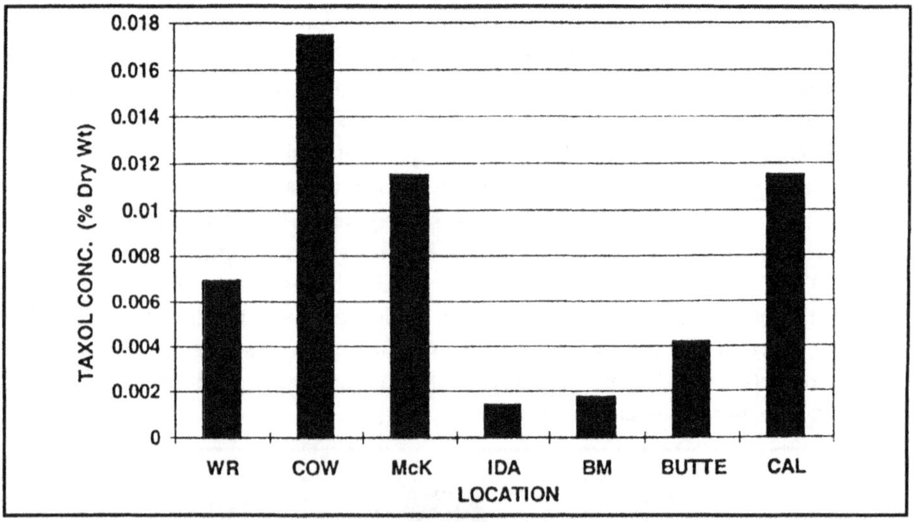

Figure 2. Variation of Taxol Content of *T. brevifolia* by Location

These samples were all collected in September and demonstrated high variability from specimen to specimen. A similar variability was seen when samples were taken from the same tree over a period of time. The vast majority of the genetic potential of the plant is silent most of the time and is expressed only under the appropriate climatic, herbivore or infectious challenge. Thus the effect of a typical phytochemical examination of a plant from single collection represents the result of a "phytochemical snapshot" of the particular species. It represents only the result of those genes being expressed under the conditions that the plant has experienced just prior to collection. Phytera's research strategy with respect to phytochemical secondary metabolites attempts to address this selectivity by establishing a "phytochemical photo album" of the biosynthetic capacity of the species under a variety of challenges.

Plants can be taken in cell cultures from virtually any live tissue including seeds, roots and shoots. Each cell has within it the full genetic potential of the plant and each gene can be expressed in response to the appropriate stimulus. Initially callus cultures are established on agar and when these have developed into a mass of about 2g they are taken into suspension culture. The suspension cultures are then subjected to a series of treatment protocols to activate a variety of secondary metabolic pathways. These protocols comprise a mix of treatments that include various hormones, elicitors, infection mimics precursor feedings, modified substrate additions and "shotgun" gene derepression. HPLC profiles of similar extracts of the same cell line routinely show the more complex nature of small molecule extractives from a manipulated cell line when compared to an untreated cell line. In Figure 3 the HPLC profiles of two extracts of different lipophilicity are shown for a manipulated culture and can be compared with two similar extracts from the native plant.

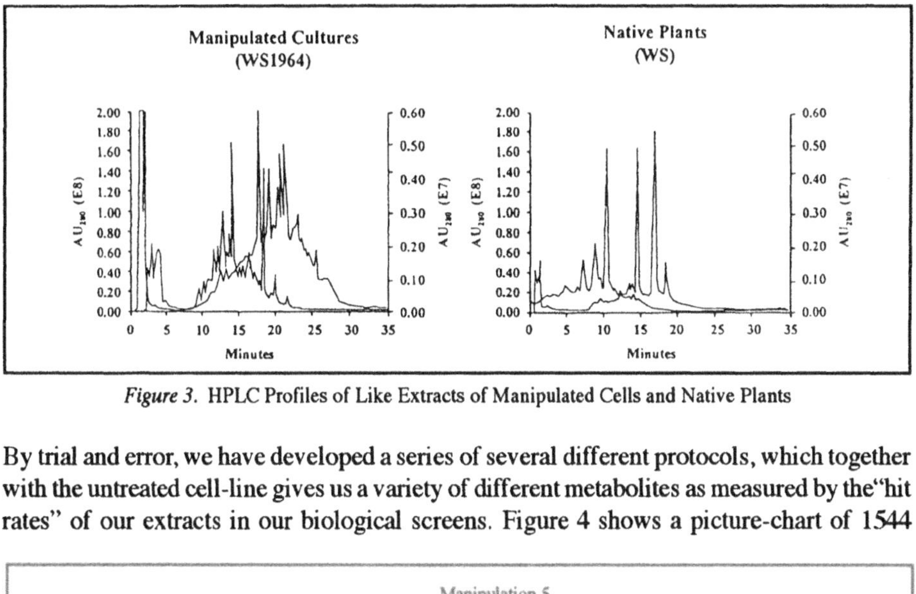

Figure 3. HPLC Profiles of Like Extracts of Manipulated Cells and Native Plants

By trial and error, we have developed a series of several different protocols, which together with the untreated cell-line gives us a variety of different metabolites as measured by the "hit rates" of our extracts in our biological screens. Figure 4 shows a picture-chart of 1544

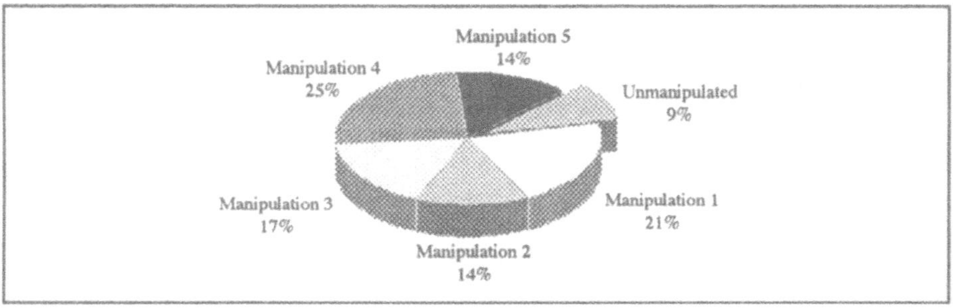

Figure 4. Contributions of Manipulation Protocols to Bioactivity of Extracts

positive extracts by treatment protocol source for five such protocols. It is apparent that all five treatments contribute substantially equally to the hits and each is more productive than the parent cell-line. In this analysis 673 "hits" were exclusive to a single treatment and most of the remainder occurred only in extracts from two treatment protocols. The Phytera collection has been drawn from the full spectrum of the higher plant orders and over 75% of described families are represented.

An important feature of our strategy is the ability to store cell lines and to be able to regenerate extracts with the same bioactive components both at some future date and in increased scale. Unlike the classical natural products approach a plant cell culture source requires very little biomass to begin with and should have no requirement for reaccess of the species in the wild. Hence there is no environmental impact nor is there the uncertainty which can occur in trying to reaccess a rare species, particularly those of a vanishing rain forest. Several methods of cell line maintenance are employed.All cell lines are cryopreserved and for those with demonstrated successful revival some months later this is the preferred storage method. The cell line may also be maintained in various growth modes by successive reculture, depending on the protocol, every few weeks to once every six months. Alternatively, calluses may be maintained in cool storage with appropriate subculture almost indefinitely. HPLC profiles of a standard lipophilic extract (as one rich in small molecule secondary metabolites) have been used to substantiate the following premises:

i.) that a suspension culture on regrowth under the same protocol gives the same profile of extractives. (Figure 5)

ii.) that the same "shotgun" protocol of genetic manipulation gives transformed cell lines which yield similar metabolite profiles. (Figure 6)

iii.) that storage of a suspension culture does not effect the ability of the genetic manipulation protocol to regenerate a cell line with a similar metabolite profile. (Figure 7)

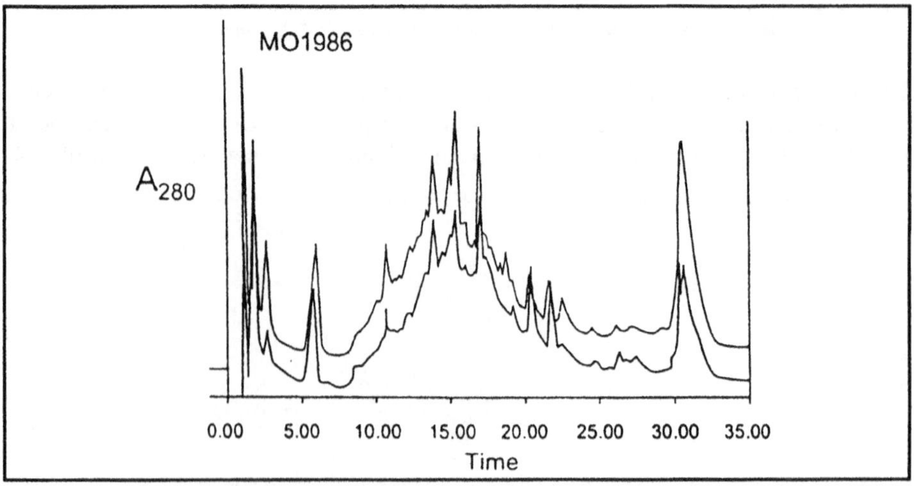

Figure 5. HPLC Profiles of Extracts from Regrowths of the Same Cell Line

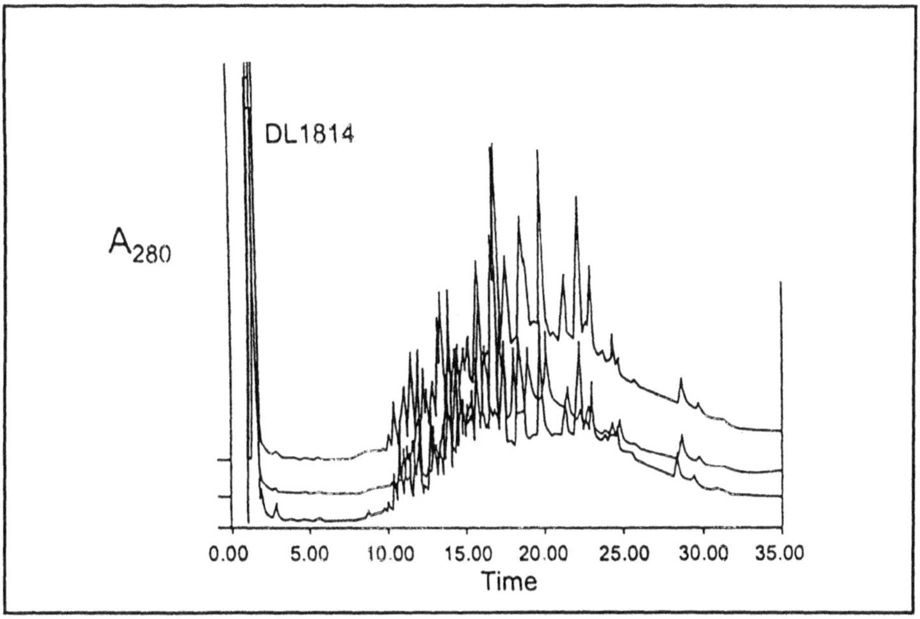

Figure 6. HPLC Profiles of Extracts from Similar Epigenetic Cell Lines

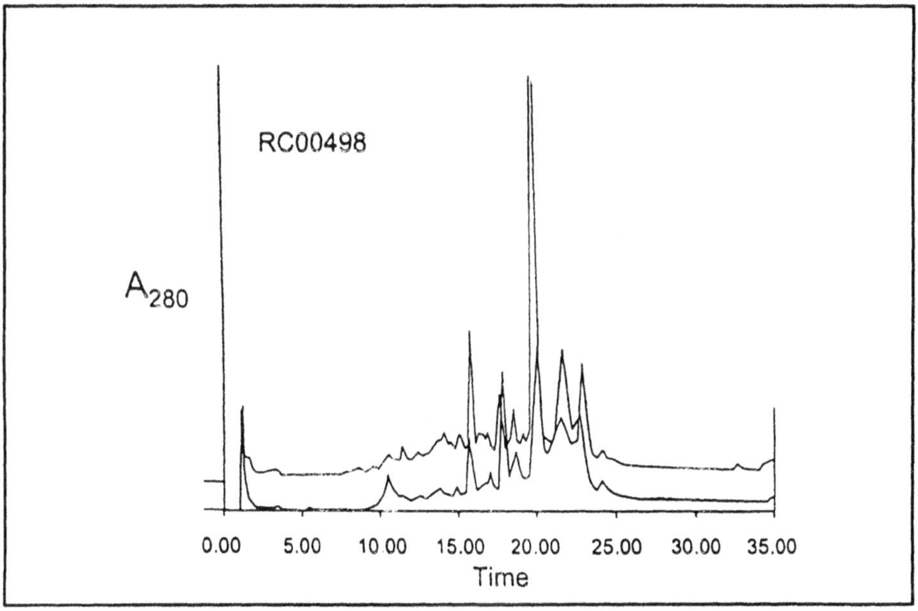

Figure 7. HPLC Profiles from Extracts of Two Epigenetic Cell Lines Generated at Different Times

At harvest the biomass is filtered from the media and a series of proprietary extraction protocols have been developed to give four extracts of varying lipophilicity and effectively concentrating derived chemical classes while eliminating those which are likely to interfere with assays such as tannins and other polyphenolics and dilutants such as fats, waxes and simple sugars.

Phytera has developed a series of mechanism based and non-specific whole-cell anti-infective assays aimed at finding antivirals, antibacterials and antifungals. Mechanistic assays have the advantage of simplifying dereplication procedures and allow for the discovery of pharmacophores which act at intracellular targets but which are unable to penetrate the cell wall. These then can form the basis for a chemical modification or synthetic program aimed at overcoming the penetration problem. Whole-cell assays provide a formidable dereplication problem but they also hold the promise of discovering an agent which acts at a new, as yet unidentified target.

In the course of screening our extracts from plant cell cultures we have discovered a novel antifungal compound we have called "Sunillin". It is a small secondary metabolite with a novel structure and with modest in vitro antifungal potency but which has shown impressive in vivo efficacy in a stringent model of mouse candidiasis run by Dr. Alan Sugar of Boston University. In this model immune competent mice are infected intravenously with *Candida albicans* and the infection is allowed to disseminate for 24 hours before treatment is initiated. At this point sacrificed animals have a general systemic infection with large numbers of colony forming units recoverable from the kidneys, liver and spleen. Treatment is administered, either by oral gavage or intraperitoneally, once or twice a day for 10 days. Amphotericin B, intraperitoneally, at 3 mg/kg, once a day is curative in this model. This is also reflected in sterile organs of sacrificed amphotericin B treated animals at the end of a 21 day experiment. Untreated infected animals begin dying after a

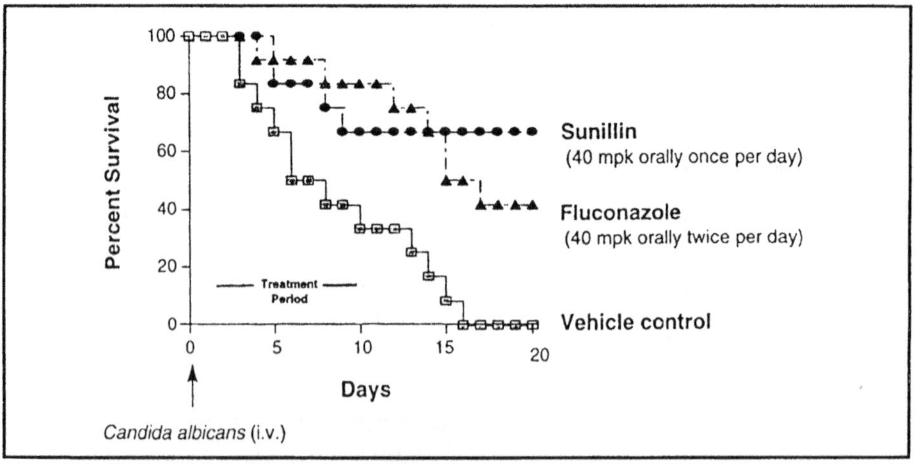

Figure 8. Efficacy of Sunillin Compared to Fluconazole in a Murine Candidiasis Model (Fluconazole Sensitive *C. albicans*)

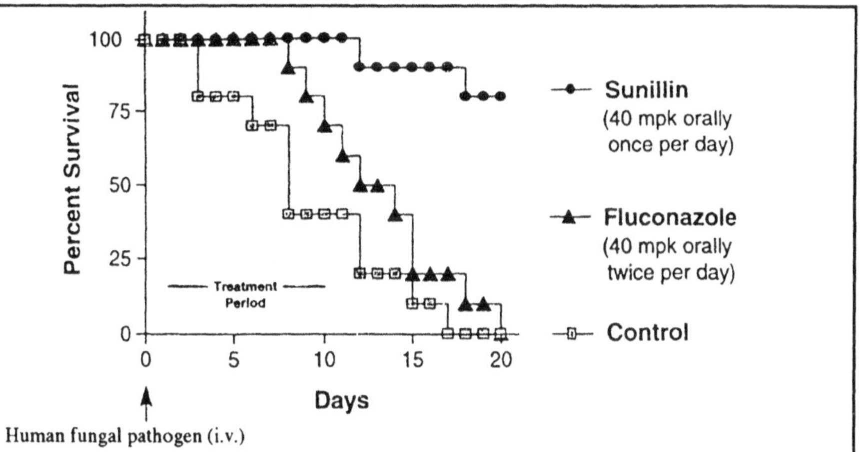

Figure 9. Efficacy of Sunillin Compared to Fluconazole in a Murine Candidiasis Model (Fluconazole Resistant *C. albicans* - Clinical Isolate)

few days and are almost always all dead by the end of the experiment. Fluconazole, the current leading drug for systemic candidiasis, when given at 40 mg/kg twice a day is clearly efficacious, but not curative in this mode. When Sunillin is given orally in PEG400 at 40 mg/kg only once a day it has comparable and usually superior efficacy to Fluconazole administered at twice the frequency if the Candida strain is Fluconazole sensitive (Figure 8) and clearly more efficacious with a Fluconazole resistant strain (Figure 9).

Of particular importance to this article is not so much the characteristics of Sunillin as an antifungal but its validification of Phytera's strategy. Sunillin was produced only by

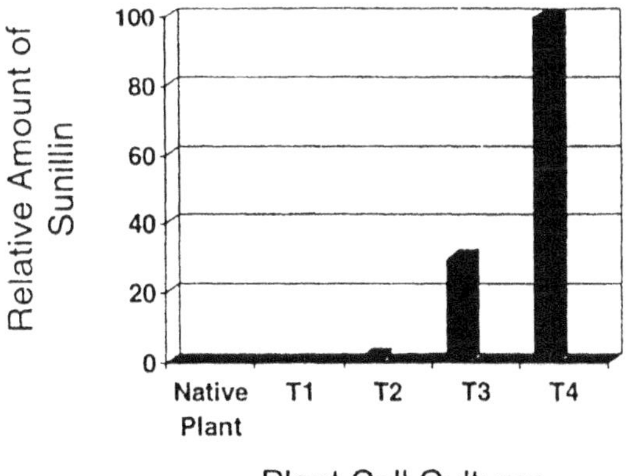

Figure 10. Abundance of Sunillin in the Native Plant and a Cell Line Under Various Treatment Protocols.

cells which had been treated in a manner to mimic the effects of a fungal infection. It was not found in the unmanipulated parent cell lime, nor in other manipulations of the cell line. Figure 10. In a number of searches it has not been found in the native plant, even though one of those followed an attempt to induce a fungal infection in the live plant.

It is clear that plant cell cultures and their manipulation, offer a unique opportunity to access a large amount of the biosynthetic potential of the plant kingdom which might otherwise go untapped.

References

[1]Cragg, G.M., Newman, D.J. and Snader, K.M. (1997) Natural Products in Drug Discovery and Development, *J. Nat. Products*, **60**, pp52-60.

[2]Wheeler, N.C., Jech, K., Masters, S., Brobst, S.W., Alvarado, A.B., Hoover, A.J. and Snader, K.M. (1991) Effects of Genetic, Epigenetic and Environmental Factors on Taxol Content in *Taxus brevifolia* and Related Species, *J. Nat. Products*, **55**, pp432-440.

EXPEDITING DRUG DEVELOPMENT THROUGH THE USE OF PLANTS

PETER J HYLANDS
Oxford Natural Products PLC, No. 1 St Giles, Oxford OX1 3JS, United Kingdom and Department of Pharmacy, King's College London, Manresa Road, London SW3 6LX, United Kingdom

1. Introduction

Drug discovery and development is long and costly. The average new time taken to put a new drug on the market is estimated to be 7-10 years at a cost which can rise to US$ 100 million. This figure itself can rise to several times that number if the costs of candidates which do not complete the development process are taken into consideration. Although it is impossible to give exact figures, it has been reliably estimated that the chance for a single drug to complete the whole process from the research idea to the marketplace is about one in ten thousand. This article looks at the causes of failure in the development process and attempts to give an account of some of the ways in which the use of plants either as leads or as a source of the drug material itself can expedite this process.

2. The place of plants in drug discovery and development

There are two principal ways in which plants have a place in drug discovery and development:

- as a source of biodiversity (hence chemical diversity) in screening programmes, and
- from the study of ethnobotany.

Each approach has its protagonists and each has had some success in leading to the introduction of plant-derived drugs into current medical practice. This paper will consider ways of expediting drug development from both points of view. Before a detailed consideration of these, however, it will be useful to consider the whole process of drug discovery and development, particularly the latter, which is often omitted from discussions of pharmaceutical research (as opposed to development).

3. The drug development process

The overall scheme of drug development from the original research idea to reaching the market is shown in Figure 1, modified from the excellent account of Yevich (1991). The process outlined is a generalised form of that in use in the United States (with the widely-accepted US terminology) but the regulatory procedures used in other countries (notably Europe and Japan) are broadly similar and follow much the same steps though individual requirements may vary, particularly in the clinical stages.

3.1. THE RESEARCH HYPOTHESIS

The initial research concept may result from the isolation of a new target (gene, receptor or enzyme, for example, or a combination of these) and the hypothesis is set up that if a compound can be discovered which modifies the action or effect of this target, then it may be possible to use this molecule in therapeutic or preventative medicine. As targets themselves become more sophisticated it may be increasingly difficult to predict the precise effect which an 'active' molecule may have or even into which therapeutic category any successful drug output may fall. Certainly there are a number of examples where a drug, originally highlighted in a search for a compound useful in one disease, eventually reaches the market in a quite unexpected and maybe apparently unrelated condition as a result of observations made during development.

3.2. SCREENING

This hypothesis is then tested by first designing an appropriate test (assay) system and subsequently exposing as wide an array of potentially active molecules to this system in order to select those which have activity against the target. This array may comprise synthetic molecules, natural product extracts, or may include mixtures of molecules produced by combinatorial methods. Current approaches are attempting to combine certain advantages of the combinatorial approach with natural product templates. An important feature of current assay methods is their miniaturised format (sometimes including so-called nanotechnology) allowing high throughput screening during which many thousands (sometimes tens of thousands) of samples are tested in minimal time. Sophisticated computerised recording and data tracking systems are obviously a prerequisite for the successful implementation of such a strategy. Methods applied particularly to natural product research have recently been reviewed (Claeson and Bohlin, 1997).

Selection criteria for good natural product libraries are discussed later but, in general terms, the more diverse the biological sources for the preparation of the extracts, the more diverse their contained chemistry and consequently the greater chances of finding *novel* chemical templates - vital for effective protection of intellectual property, a major criterion in the selection of potential clinical candidates.

3.3. LEAD OPTIMISATION

Once a chemical **lead** has been initially identified by this process, the usual next step is the initiation of a programme of analogue synthesis in order to elucidate structure-activity relationships within the chemical series. A goal of this phase of the process is the preparation of a better **development candidate** which has more desirable properties such as improved efficacy or reduced undesirable side effects or toxicity. In short, this stage is directed towards the optimisation of the therapeutic index, *i.e*, the ratio of drug dose which causes undesirable side effects or toxicity to that which produces the desired therapeutic response. The role of computer assisted drug design methods cannot be overemphasised at this stage in order to minimise the number of analogues to be synthesised. This process is becoming simplified as data retrieval systems increase in sophistication and effectiveness. A small group of potential development candidates are likely to be selected from this phase and they next pass to experimental pharmacology in order to assess more completely their pharmacological profile. A candidate is selected from these studies at the completion of this optimisation stage as having real potential for clinical evaluation. However, the process begins now to divert from a linear track to involve a number of parallel activities designed to prepare the compound for the first administration to man.

3.4. PREPARATION FOR IND FILING - PREPARATION OF PRE-CLINICAL DATA PACKAGE

In order to begin the clinical evaluation process, the regulatory authorities must be satisfied not only with regard to the clinical protocols to be followed but also that the putative drug has been properly evaluated in animal toxicity studies and that its metabolic and pharmacokinetic parameters are adequately understood. From the manufacturer's point of view, it must be certain of being able to produce material for clinical evaluation which is sufficiently pure and also a formulation suitable for the delivery of therapeutic levels of the drug. All these studies must be undertaken more or less simultaneously with considerable attention to logistics and planning to minimise the time (and hence cost) taken. However effective the planning, costs begin to rise significantly at this stage.

The following activities are thus involved: the company's chemical process development group must develop appropriate synthetic methods (or make other special arrangements in the case of natural products such as fermentation or pilot plant extract groups) in order to make available bulk (kilogram) quantities of the raw drug substance. For synthetic compounds, this may well require the development of entirely new chemical methods since, at the pilot (and larger) scale, the number of operations such as chromatographic separations must be minimised or their use obviated entirely in order to develop a commercially viable process. Such considerations are especially important in the development of processes involving complex mixtures such as plant extracts or fermentation broths. A further important issue that drug materials for human consumption must be prepared ac-

Research phase (1 - 3 years)

Discovery paradigm

↓ Screening

Nomination of lead template

↓ Lead optimisation

Identification of a 'clinical candidate'

↓

Preclinical development phase (1 - 3 years)

↓ Process development, pharmacology, formulation, subacute and acute animal toxicology, metabolism and pharmacokinetics

Submission of 'IND package'

↓

Clinical development phase (3 - 6 years)

↓ Phase I - III clinical evaluation, human and further animal metabolism and pharmacokinetics, chronic toxicology

Submission of "NDA package'

↓

Regulatory approval (~2 years)

↓

Marketing

Figure 1 - Schematic showing in simplified form the stages in drug discovery and development. The times indicated are rough generalisations designed to show the broad differences in time (and hence cost) of the research and development stages.

cording to the guidelines of good manufacturing practice (GMP) which relate not only to standards of operation (particularly record keeping which must be comprehensive and scrupulous) but also to the choice of reagents themselves.

The test materials so produced are then formulated by the pharmaceutical development group into forms suitable for clinical evaluation, capable of releasing sufficient of the drug to allow it to exert its desired effect. The dosage form selected must also obviously be sufficiently stable to ensure reproducibility. The drug material itself is studied with increasingly detailed pharmacological tests to understand the compound's wider biological activity. Also, the material is studied in animals to determine the compound's metabolic fate (including the production of metabolites) and its pharmacokinetics (criteria establishing absorption, distribution and elimination). Bioavailability studies are particularly important because a later discovered lack of clinical response with a compound which shows excellent *in-vitro* behaviour may be due to poor absorption, rapid and/or complete elimination or sometimes metabolism to inactive metabolites. Active metabolites or those which possess undesirable effects not shown by the original substance may also be produced. Alternatively, the active drug may become strongly bound to plasma or tissue proteins, effectively delaying or preventing therapeutically effective circulating levels from being achieved. If such properties are discovered it may be possible to synthesise analogues or pro-drugs which solve these problems. Some metabolic studies may be undertaken with *in-vitro* preparations or liver enzymes or cells but some are also carried out in whole animals using isotopically-labelled (radioactive or stable) materials. Potential *in-vitro* drug interactions may also be examined at this stage using newly developed methods.

Equally important, the toxicology of the compound is examined. The usual first toxicity evaluation examines the putative drug's potential to cause genetic damage or mutation. Such tests are carried out using specifically-modified strains of micro-organism such as *Salmonella* in the Ames test. Compounds which do not induce mutagenesis are then tested for any short term side effects initiated by a single dose in whole animals, using two mammalian species, usually in one rodent and one non-rodent. The choice of animal species is as far as possible guided by their similarity to man with regard to pharmacokinetics including biotransformation. The behaviour of the animals is closely observed as well as their blood biochemistry, liver and kidney function. At the end of the study, the animals are killed and their organs examined closely for drug-induced abnormalities. Range finding studies are then undertaken in which animals, again usually both a rodent and non-rodent, are repeatedly dosed in order to establish the correct dosing regimens for later subchronic (studies 1-3 months in duration) designed to study adverse reactions produced by repeated administration. Chronic studies, usually 6-12 months in rat and another species, determine the long term toxicological effects of the drug. Special toxicity studies are expressly designed to evaluate the drug's effects on reproduction and its potential to induce embryotoxicity and teratogenicity. A particular form of chronic evaluation is an 18-24 month carcinogenicity study designed to evaluate the potential of the substance to induce the development of tumours on extended administration. Carcinogenicity evaluation requirements depend on the intended route of administration, structure, mechanism of action and the results of earlier findings.Indeed, the precise protocols for the entire toxicological evaluation depend on the type and nature of the indication of the intended use of the drug.

The observation of any untoward findings in the above stages such as unexpected toxicity, difficulties with producing a stable formulation, or poor pharmacokinetics, for example, will result in the cessation of the project or at least the initiation of the search for a better candidate from within the chemical series.

Assuming that the compound shows satisfactory behaviour in the above studies, the sponsoring company will initiate a request to the regulatory authorities (the Food and Drug Administration [FDA], in the United States) in the form of an Investigational New Drug (IND) application to begin evaluation of the substance in man. The package includes data generated in the pre-clinical studies above as well as details of the proposed clinical studies including details of the human protocols. The whole data are then reviewed by a panel of independent scientists, clinicians and laymen.

3.5. CLINICAL STUDIES

3.5.1. *Phase I*

The clinical studies are the most time consuming (and costly) stages of drug development. They usually begin with evaluation of the putative new drug in healthy volunteers (not patients) except in certain special conditions such as cancer and AIDS. The object of these studies is to establish the compound's safety in man and to establish suitable dosage levels for the next phases. The drug is administered first as a single dose and the subjects (usually 20 - 100 young men) are observed closely, records being made of temperature, blood pressure, pulse, changes in blood and urine chemistry as well as any reported perceived side effects. Study of blood and urine samples gives information about the drug's human pharmacokinetic parameters. Subsequently the studies are extended by the administration of repeated doses of up to 3 - 4 times daily, depending on the properties of the drug. Assuming that no untoward side effects are observed, the next stage in clinical evaluation is the first time that the drug has been tried on patients suffering from the condition which the drug was designed to treat - phase II.

3.5.2. *Phase II*

Phase II studies determine appropriate dosing regimens and whether the drug is effective in treating patients suffering from the intended therapeutic target. These trials typically involve several hundred patients, often at a number of sites. The dose initially deduced from the phase I volunteer studies is usually modified in order to achieve maximal efficacy with minimal side effects. This is conveniently achieved in open studies (in which all the patients knowingly receive the new drug). True efficacy studies are carried out in some form of blind test, usually randomised, in comparative studies. In such tests, the efficacy of the drug in question is compared with the effects of either a known, existing treatment or an inactive placebo substance, or both (in multiple-arm studies). More complex trial designs involve a cross-over phase, in which a randomised group of patients receive both treatments successively. In double blind studies, neither the patient nor the administering physician knows which treatment the patient is receiving. Such studies are the subject of rigorous statistical controls and analysis. For example, those involved in the preparation of the test materials are not involved in the later clinical evaluation stages.

3.5.3. *Phase III*

Phase III extends the clinical evaluation to wider use in order to verify efficacy and monitor possible adverse reactions from administration over more extended periods, as well studying potential drug interactions and the possibility of complications arising in patients suffering from unrelated illnesses. Typically, these studies can involve several thousand patients and last for a number of years. If the data show that the drug is effective and safe and is likely to form a contribution to the therapeutic armamentarium, they are submitted to the regulatory authorities in the form of a New Drug Application (NDA).

3.4. FILING FOR APPROVAL TO MARKET - THE NDA APPLICATION

A scrupulous review process by the regulators of the copious documentation submitted by the manufacturer of the potential new drug follows which can take as much as two years. In successful cases the result is the issuance of the approval - in the form of a marketing authorisation - required to begin sales of the new drug. Clinical studies continue even at this stage however. The so-called Phase IV generates a larger data base and may highlight adverse reactions of such low incidence that they could not have been detected in the more restricted phases I - III.

4. The chances of success

In broad terms, only one in four putative new drugs survive the clinical phases to reach the NDA application. Although about 70% of candidates pass successfully from phase I to II (those which fail usually show some unexpected side effects), two in three of those which do enter phase II fail then, usually because of lack of demonstrated efficacy. The individual costs associated with a trial may be of the order of several thousand dollars per patient. Phase III studies typically involve several thousand patients so it is obvious that costs rise dramatically as the development process continues. The relative costs associated with the process are shown diagramatically in Figure 2, from which it is clear that the most significant contribution to cost is associated with the clinical phases.

It follows from the above that any changes which can be made to reduce the drop out rate so that resources and effort may be concentrated on projects which are destined to be successful, are likely to be cost effective and increase the overall benefits of success. The various ways in which this may be accomplished by the use of plants are discussed below.

5. Criteria for success

It is worth examining those parameters which have to be addressed for successful drug development. Apart from the obvious aspects such as efficacy and safety, there are, among others, three principal criteria which ideally should be met:

- a strong intellectual property position,
- the compound should serve an unmet clinical need, and
- the compound should operate by a novel mechanism of action.

There are a number of reasons, scientific, commercial and regulatory, which fuel these criteria. The first concerns the ability of a company to protect its intellectual property. As has been seen, the costs of drug development are so high that no corporation could entertain the notion of developing a product over which it did not have some strong level of ownership. Usually, this is available through the use of patent rights, which allow exclusivity in the manufacture and sale of the drug so that the company can recoup its research and development costs and generate profit to invest in further projects. The simplest and most straightforward patent is the so-called composition of matter patent whereby a specific compound (or usually a series of structurally related analogues) is protected for some specified use, sometimes along with methods of preparation. Such patents may be granted only where the compound concerned has not previously been reported in the literature. Such patent protection is usually clear and unequivocal to prosecute and as such is most highly valued as a contribution to a company's intellectual property portfolio. Any action to increase the chances of finding novel chemistry as a result of a screening programme is thus very important.

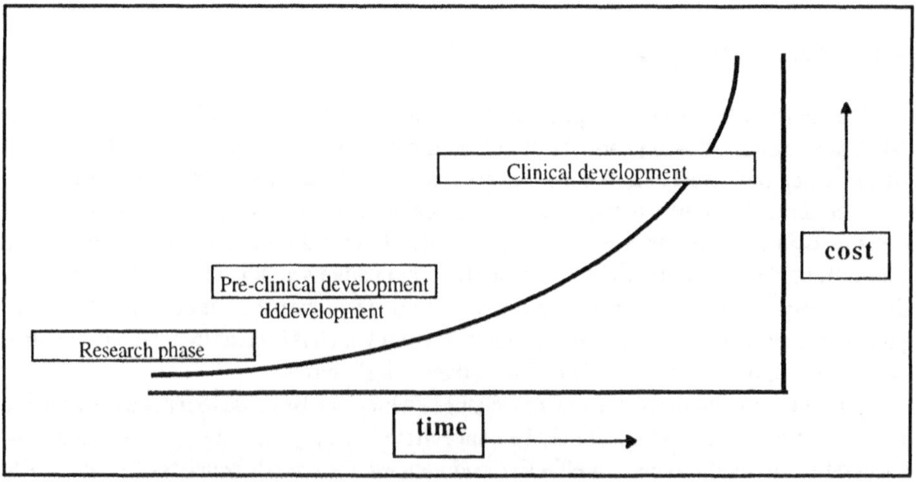

Figure 2 - Schematic showing how cost rises during the discovery and development process.

Other types of patent are 'use' and 'process' patents. The former may allow some degree of exclusivity by being granted on a known chemical (and therefore not able to be protected by a composition of matter patent) for a use neither previously reported nor expected from the available literature. Process patents may be allowed for new or improved processes, often those used in commercial production.

Unless an alternative process which is as effective as the patented one is developed by a competitor, the latter may preclude entry to the market by another company, so effectively increasing and extending the intellectual property position of the originating organisation.

The second criterion relates to the potential compound's finding a significant role on the marketplace. Not only must the drug be effective and safe but it must also have a chance of generating sufficient sales income to justify the original investment. This commercial consideration sometimes has the effect of eliminating certain minor conditions as targets by the industry, particularly for the very large companies for whom annual sales figures in the region of $100 million may be the minimum for a viable project.

The third criterion is imposed to a large extent by the regulatory authorities who have an increasing reluctance to approve for sale those drugs which belong to known chemical classes operating by a well-understood mechanism of action and which may thus show only marginal therapeutic or toxicological advantage over existing products. A distinct advantage over existing therapy is thus, not only a commercial (marketing) matter but almost a regulatory prerequisite.

6. **How the use of plants may expedite the process**

6.1. THE ROLE IN SCREENING

Other contributions to this book are concerned with screening approaches but it worth mentioning again here to emphasise methods which are designed to maximise the chemical diversity fed into screening programmes. It is well known that whereas about one in four prescription medicines are derived directly or indirectly from natural sources, less than 10% of plants (and the figure is much less for microbes) have ever been examined even in the most cursory way for their pharmacological potential. Since novel chemicals are required, it follows that previously-unexamined sources should be the highest priority for inclusion in natural product libraries. This requires a reliable knowledge of the identity of the original materials combined with exhaustive use of databases to prioritise those samples which have not been studied previously. It is a pity that negative screening results are not published by those organisations who carry out such work. The generation and use of in-house databases is of great importance in the industry and they provide much valuable information about selectivity cross-reactivity. Of course, this does not mean that only 'new' sources should be screened. New assay systems are constantly being discovered and developed and extracts which have shown no previous activity may well contain molecules active against the new target (for an account of the newly-accepted role of pharmacognosy in such studies, see Bruhn and Bohlin, 1997).

Furthermore, the tendency should be for the treatment of samples as *individuals* rather than examples of previously published taxa.This approach takes into account the inter-specific variation discussed elsewhere in this publication as well as the local genetic variation becoming more and more apparent as detailed analytical studies are carried out in different specific populations. Moreover, the existence of highly localised (endophytic, for example) microbial/plant interactions is being recognised as an important source of novel metabolites. Approaches to optimise for these criteria should be emphasised in techniques for the provision of library samples. The geographical origin of the samples can be well documented using Global Positioning System (GPS) technology which can readily locate collection sites with a precision of a few metres. Using this method, single specimens of large trees and shrubs can be individually marked so ensuring precise recollection of the original source if the original extract showed desired activity. Such precision also allows ready relocation of collection sites within a defined population, a particular stand of shrub, for example, or where activity may be due to a plant/insect or plant/microbe interaction.

Other contributions to the publication also highlight the use of plant cell tissue culture both in the production of secondary metabolites but also to generate chemical diversity by the use of, for example, chemical and other elicitation techniques.

6.2. ENHANCING THE VALUE OF ETHNOBOTANY

As is apparent from the research and development outline above, any change in the process to reduce the chances of drop out in the later stages of development would have a significant impact on the cost of the overall process. There are two possible ways which may be considered as expediting this process:

- the improved application of the findings of the study of traditional medicine to pharmaceutical development, and
- the potential of the use of plant extracts as pharmaceutical products themselves rather than as a leads for subsequent research.

Both these approaches highlight particular problems and deviations from the orthodox research and development activities outlined above, and are discussed in more detail in turn below.

6.2.1. *The rational application of traditional medicine findings*

Traditional medicine has historically served well in discovery of potential new drugs but one of the major contemporary criticisms of its continued application is the often poor quality of the evidence, of necessity anecdotal, for the use of the plant. These combine to produce a situation where traditional medicine is perceived by many in the pharmaceutical industry as a poor predictor of successful drug development. Two main causes of particular concern are correct diagnosis and unequivocal identification of the plant. Without quantifiable data, it is difficult to extrapolate the apparent findings of traditional medicine studies to the orthodox clinic. One way round this problem is to carry out limited but *properly designed and executed* clinical trials on the traditional medicine itself before transferring the studies to the laboratory.

The identity of the plant material tested can be assured by close collaboration between the investigating centre (research institute, university department or hospital) and appropriately trained field and systematic botanists working with the traditional healers. Appropriate pharmaceutical studies can be undertaken locally to ensure that the plant material to be tested is prepared consistently in a controlled way even if true standardisation is impossible because of lack of information about the identity of the active constituents. Essentially, this would lead to the preparation of a detailed monograph on the plant material and a precise protocol for the preparation of the final dosage form. Concerns about diagnosis can be allayed by ensuring that the recruiting (conventionally-trained) physicians use, as far as possible, contemporary diagnostic methods such as X-ray analysis, blood biochemical and histopathological examinations, etc. These parameters should be as quantifiable as possible in order to increase the generation of objective rather than subjective data from the later clinical studies. The design of the trial can be as far as possible as that for such a study carried out in the orthodox western examination of a potential new chemical entity in that condition. Thus the trial design could be two- or even three-armed, one group of patients receiving the material under test, another a placebo and another a known active (if one exists for the condition being investigated). Randomisation or patient matching should be rigorous and informed patient consent should be obtained. Much attention should be paid to the selection of patients for the study. Particular emphasis should be paid to inclusion, exclusion and withdrawal criteria. For example, trials involving extended administration of material about which little toxicological data is available should exclude as subjects children and those women likely to become pregnant. Regulatory approval is usually readily obtained in the originating country because of the long local history of use of the plant material for that condition by the traditional healers of that culture which has already provided a body of evidence concerning efficacy and non-toxicity (otherwise the plant would not be the subject of the study in the first place!). If considered necessary, limited toxicological examination could be carried out on the test preparation before the initiation of the study. Patients can be properly followed up and the occurrence of adverse reactions monitored and recorded.

If the preparations for the study are scrupulous (good and unequivocal identification of test material, the use of established inclusion and exclusion criteria for patient recruitment, statistically valid randomisation criteria and high quality record keeping, excellent clinical and laboratory (if appropriate) examination during the monitoring phase), and in particular the conduct of the trial is properly controlled and evaluated, the findings can be accepted with a high level of authority and can so serve as a good guide for further studies. Real efficacy assessment combined with preliminary toxicological investigations can so be obtained with the added advantage that the overall costs involved are minimal compared with orthodox studies. There is of course no suggestion that clinical data so generated form part of the package submitted for regulatory approval but a positive finding can serve to increase the level of confidence by the researchers in continuing the study to the more orthodox pharmaceutical phases beginning with a chemical and pharmacological evaluation of the plant medicine. Such an approach has the advantage of asking the critical question (does the medicine work in the clinic?) at an early stage in the research and development process, *i.e*, it brings forward to an early stage of the discovery and development process one of the most significant hurdles, one at which it is known that some 66% of drug candidates fail. The trials carried out in originating countries cannot be carried out to the standards required by Good Clinical Practice (GCP) and so could not form part of an approval package but rather to act as a good guide for the subsequent selection of research projects.

In this way, the data are effectively transformed from uncontrolled (essentially anecdotal) case histories, and hence scientifically unconvincing, into high quality, reliable information which can at least assist in the prioritisation of research projects in the mainstream industry.

Furthermore, it may be possible to use the data generated in such studies to support an application to carry out analogous studies in western countries. One way which this approach has found application in the developed world is exemplified by some of the earliest work on the studies designed to investigate the efficacy of feverfew leaves in the prophylaxis of migraine headaches in the early 1980s (Johnson *et al.*, 1985 and Hylands *et al.*, 1985). The clinical approach here was the use of *already self-medicating* patients (the plant was widely available) and designing a trial in which those patients were transferred randomly either to a placebo or a controlled preparation containing the active. Effectively, therefore, this was a process of *controlled withdrawal* of medication. This successful study allowed progression to more conventional trials in patients who had not previously taken the plant.

However, it must be stated that this approach does not address the issue of the other parameters enumerated above, namely, novelty of chemistry and mechanism of action, prerequisites, as has been seen, for the successful development of a viable new drug entity commercially. Thus, although a new clinically effective plant extract may result from the study, more detailed pharmaceutical research and development may show that the active compound or compounds are well known or from a well studied class (with resultant poor or non-existent intellectual property protection) or that the active lacks potency and/or selectivity and so is not a good clinical development candidate. However one way to obviate these concerns has been exemplified by a recent initiative in plant medicine whereby attempts have been made to try to broaden the whole concept of pharmaceutical substances by the consideration of the medicine itself (*i.e.*, the plant extract) as a pharmaceutical material (see, for example, Guy and Whittle, 1995 and Berg, 1996). Matters which have to be taken into consideration for this approach are considered below.

6.2.2 *Plant extracts as pharmaceutical products*

At first sight, it may be considered that the registration of a crude plant extract as a pharmaceutical product is not very different from the development and registration of a new orthodox pharmaceutical. There are however a number of problems with the approach, commercially and regulatory, which are discussed below.

The first issue concerns intellectual property. As has been seen, a pharmaceutical company usually requires a very strong patent position on a compound before embarking on the costly development track. It would be suicidal for a company to invest the huge resources to achieve a marketing authorisation for a new drug if a competitor were free subsequently to introduce its own version of the substance on the back of the original company's investment, as would be the case if there were no patent protection. A change in the approach to patent policy is therefore needed. In conventional allopathic medicine, intellectual property is generally protected as new chemical entities by composition of matter patents (which incidentally allows for a steady continued improvement in position by extending the cover around the chemical series as new compounds are synthesised, as well as making compliance with the chemistry and pharmacy issues relatively straightforward because only one entity is concerned). With the newly proposed plant extract based therapeutics, which uses whole plant extracts (or even mixtures of unrelated plant extracts) the approach must be one of the application of process and use patents. Goals of patenting should be, with regard to the composition: the setting of appropriate range values in profiles or different classes of constituent, and with the process: the use of appropriate (maybe even proprietary) extraction and purification technologies. Also, with the increasing availability of methods for the almost routine genetic modification of plants, it is likely that the use of genetically engineered plants, created to confer some advantage such as consistency of active material, will increase and so provide for the possibility for an increased level of intellectual property position.

The second issue concerns the data required to be submitted to the regulatory authorities as part of the application for marketing approval. Obviously, clinical efficacy data must be convincing and unequivocal because there is no doubt that no company would contemplate making an application unless it had data to prove its product's effectiveness. Indeed, if the approach had entailed the method outlined above by obtaining early clinical confirmation, this would have the effect of increasing the latter's level of confidence about pursuing more extensive clinical evaluation. Similarly, the fact that the material was an established traditional medicine (and presumably at least anecdotal evidence of lack of toxicity had already been established) would have encouraged the sponsoring company's approach in the first place, and it may in fact be possible to limit the required toxicological studies because of the medicine's presumably well established traditional use.

It is in the area of the chemistry and pharmacy data in the regulatory package that the major changes will be encountered. As has been indicated above, many data are required to confirm chemical identity, purity and consistency of the drug substance as well as information related to the absorption, distribution, metabolism and elimination of the material. Such data are not readily available where the active is an extract of a plant which is likely to contain a mixture of compounds, perhaps in a number of chemical classes, and even more difficult to generate when the traditional medicine being considered is a mixture of extracts of a number of botanically unrelated plants.

Furthermore, as more and more evidence for the existence of synergism either within an extract of a single species (rather than an isolated constituent) or for a mixture of extracts of unrelated plants becomes recognised, the route of pharmaceutical registration of extracts outlined above may be the only one which can take advantage of these effects in the provision of mainstream medicine. A new approach is obviously required to allow registration of such materials.

6.3. ADVANTAGES OF THE APPROACHES

The systematic scientific evaluation of traditional medicines in the clinical way described above may provide a sound basis for the manufacture and sale of traditional medicines in the originating (usually developing) countries. Medical communities in developing countries are sometimes resistant to the widespread introduction of traditional medicine in the form of plant extracts which is perceived as low-tech' and of doubtful value but, where they can be shown to be effective, may provide a cheap, readily accessible (and accepted) and locally produced alternative to prohibitively-expensive western medicines and so make a major contribution to primary health care. The studies which could form the basis of the introduction of these medicines are cheap and relatively quick to perform and their implementation could be facilitated by the provision of grant support by the international agencies.

The introduction of cheap and effective medicinal products in this way can also play a significant role in the provision of effective treatments for tropical diseases, research and development of which would not be undertaken by the mainstream, western pharmaceutical industry because of low return. Systematic proper evaluation of local traditional pharmacopoeias would thus encourage the development of local pharmaceutical industries. In addition, such studies could add the necessary value to the potential products before undertaking detailed research and development in order to possibly develop new mainstream drugs in the form of isolated chemical entities. This would facilitate the interaction between western pharmaceutical industries and research and development (and clinical research) institutions in developing countries.

The second major advantage centres round the fact the technologies required for successful development are disparate and readily accessible by the increasingly applied virtual approach (e.g., Spilker, 1996). Specific benefits are cost reduction and increased direct control over them, the ability to use the most respected, experience and relevant consultants and, above all, the simplicity of organisation with concomitant improvements in efficiency and communication, allowing priorities to be set and maintained at a high level (with less competition for resources) which culminate in reduced development times and cost. This approach is discussed in Davidson *et al.*, 1996 and some current examples of companies operating more or less virtually are given in Leheny and Wood, 1996.

Finally, there is a major prospective advantage in the western world. With global changes in the provision of health care resources and the obvious inevitable increased reliance on preventative measures rather than those based on late stage intervention therapy, the introduction of relatively low cost, health-promoting plant drugs has immediate economic and social benefits. This is particularly apparent with current developments on the pharmaceutical benefits of dietary components, such as those in cancer prevention. Furthermore, even a cursory glance at the scale to which commerce and trade in plants and plant compounds have grown in the United States (e.g., Brevoort, 1997) in recent years should convince even the most staid, orthodox pharmaceutical executive of the commercial benefits of the approach. The author has no doubt that this route for increasing and promoting human health will become a significant factor in the provision of worldwide health care for the third millennium.

References

Berg, J. (1996) M. H. Meyerson Research Reports, 11 November 1996.

Brevoort, P. (1997) The US botanical market - an overview, *HerbalGram* **36**, 49-57.

Bruhn, J.G. and Bohlin, L. (1997) Molecular pharmacognosy: an explanatory model, *DDT* **2**(6), 243-246.

Claeson, P. and Bohlin, L. (1997) Some aspects of bioassay methods in natural-product research aimed at drug lead discovery, *Trends in Biotechnology* **15**, 245-248.

Davidson, D., Hylands, P.J., Sharp, W.R. and Stahlhut, R.W. (1996) Development of pharmaceutical companies based on plant products: suggested approaches, in M. J. Balick, E. Elisabetsky and S. A. Laird (eds.), *Medicinal resources of the tropical forest - Biodiversity and its importance to human health*, Columbia University Press, New York, pp. 19-40.

Guy, G. and Whittle, B. (1995) Traditional plant medicines - a new approach, *Pharmaceutical Biotechnology International*, 23-26.

Hylands, D.M., Hylands, P.J., Johnson E.S., Kadam, N.P. and MacRae, K.D. (1985) Efficacy of feverfew as a prophylactic treatment for migraine, *The British Medical Journal* **291**(6508), 1128.

Johnson, E.S., Kadam, N.P., Hylands, D.M. and Hylands, P.J. (1985) The efficacy of powdered dried leaves of *Tanacetum parthenium* (L) Schultz Bip. in comparison with placebo as a prophylactic treatment for migraine, *The British Medical Journal* **291**(6495), 569-573.

Leheny, A.R. and Wood, M. (1996) The New Drug Discovery Machine, Hambrecht and Quist Institutional Research Biotechnology Report, pp. 1-14, 18 September 1996.

Spilker, B. (1996) Virtual drug development on a global basis, *Drug News & Perspectives* **9** (9), 524-531.

Yevich, J.P. (1991) Drug development: from discovery to marketing, in P.Krogsgaard-Larsen and H.Bundgaard (eds.), *A textbook of drug design and development*, Harwood Academic publishers, Chur, pp. 607-630.

THE FUTURE ROLE OF NATURAL PRODUCTS IN DRUG DISCOVERY

ANTONY D. BUSS
Bioprocessing Research Unit
Glaxo Wellcome Research and Development
Medicines Research Centre
Gunnels Wood Road
Stevenage
Hertfordshire SG1 2NY
UK

1. Introduction

The healthcare industry in facing continued pressure to bring innovative new medicines to market faster, more efficiently, than ever before. The drug discovery process must change to meet these demands and most companies have been eager to grasp the new technologies associated with automated high throughput screening and combinatorial chemistry with the aim of identifying rapidly more lead compounds for development. One aspect of the problem was featured in the Wall Street Journal of 12 August 1997 which highlighted the fact that 40 drugs with annual sales of $16 billion are set to lose patent protection in the next 5 years. Never before have so many important drugs come off patent in such a short period of time and the major pharmaceutical companies are racing to find new drugs to replace the billions of dollars in sales they stand to lose.

In the past, screening capacity outstripped the number of available synthetic compounds and natural product extracts filled the gap (for Glaxo Wellcome this represented up to 75% of samples screened). Screens sometimes ran for several years which allowed plenty of time to follow-up and characterise active plant or microbial samples. However, all this has changed. Rapid advances in molecular biology, genetics and increased understanding of disease mechanisms have led to many more novel targets being developed into sophisticated high throughput screens. It is not unreasonable for a company to now screen against 50 or 60 targets per year with a range of sample types; there are hundreds of thousands of synthetic compounds available to purchase and combinatorial chemistry offers the chance to generate many more [1]. This, coupled with much higher throughputs made possible with robotics and miniaturised screening formats [2], means that >150,000 samples can be screened against a target in 2-3 weeks. The most likely constraint will be the cost!

2. Natural Products in Drug Discovery

The past success of natural products in the drug discovery process is clear; over half today's best selling pharmaceuticals are derived from natural products. However, past success does not mean we should continue in exactly the same way. In fact, natural products screening has become tarnished by the frequent isolation of known compounds, the undue length of time required to characterise novel actives and the difficulties in generating analogues from new, often complex, chemical templates.

Combinatorial chemistry will provide solutions to many of these problems, however, it is by no mans a perfectly developed technology. At present, the number of applicable chemical reactions is small and in particular, the availability and cost of monomer starting materials is limiting. It is difficult, for example, to imagine small molecules from current libraries exhibiting the exquisite protein-protein binding properties displayed by the microbial metabolites cyclosporin, FK506 (tacrolimus) and rapamycin. The challenge, while combinatorial chemistry technology is developed further, is to maximise the chances of new drug discovery by utilising nature's rich chemical diversity more efficiently. The ability to characterise novel natural product samples rapidly is one obvious goal.

3. Higher Quality Samples for Screening

One approach to address the problems associated with screening crude natural product extracts is to fractionate the samples *before* biological testing. At Glaxo Wellcome we have chosen to fractionate by HPLC at a scale which gives samples weighing between 1-10mg. Methanol extracts of dried plant material or microbial fermentation broth are processed to provide a desalted concentrate ready for preparative reverse phase gradient HPLC. The fractions produced are free from fatty acids, chlorophylls and tannins. They are not highly coloured and are often pure enough for NMR analysis; typically half the samples contain >50% a single component, others are more complex mixtures, whilst some are essentially pure (>90%). This strategy provides samples of good quality for screening and is also operable at high throughputs.

Preparative reverse phase HPLC fractionation of a methanol extract derived from the plant *Leucanthemum serrotina* is a typical example. Fractions were collected automatically over 1 hour at intervals of 30 seconds and subsequently, a number of the samples were found to exhibit good activity in a screen for mast cell degranulation inhibitors. The corresponding 4 groups of 9 fractions were analysed directly by NMR without the need for further purification. The active metabolites were identified as a bis-caffeic acid ester derivative, 3 eudesmodienolides and a hydroxylated isoflavone.

The advantages of the prefractionation approach are that higher quality samples are screened at normalised concentration, common non-selective metabolites are removed and individual active compounds can be isolated very rapidly.

4. Improved Fraction Collection

Further improvements to natural product isolation rates are illustrated by two simple modifications to HPLC fraction collection which allows for rapid bioassay-guided purification. A new base plate configuration built onto a Gilson 222 XL liquid handler (lower base-plate and accommodation for 6 x micronic racks) allows fractions to be collected automatically into 1ml micronic tubes and then loaded directly onto a centrifugal vacuum concentrator. The dried samples can then be dissolved into aq. DMSO for screening. Collecting 40 fractions per sample, 12 chromatographic runs can be made overnight without manual intervention.

Another modification uses a 10 x 8 microtitre plate array of solid phase extraction cartridges (each containing 50mg of 40μ C18 silica). Eluant from the detector is diluted automatically with water via an additional pump and the combined streams directed into the SPE cartridges. Metabolites bind to the silica, are washed with water and eluted with organic solvent to produce desalted samples ready for assay (HTS and/or mass spec.).

5. Chemical Diversity Beyond that Found in Nature

The initiatives described above provide significant efficiency gains for natural product research, however, there is an exciting new technology which presents a real opportunity to design and generate libraries of diverse molecules that are not accessible by chemical synthesis, nor are ever likely to be found in nature!

The technology is known as combinatorial biosynthesis and recent research at the Universities of Cambridge in the UK and Stanford in North America has uncovered a number of related biosynthetic pathways with fascinating consequences for generating new molecules. The products of these pathways are polyketides; a diverse family of natural products whose apparent structural complexity shadows an underlying common biosynthetic mechanism. The polyketides are no mere chemical curiosity, for many exhibit a wide range of important biological activities. The fact that polyketide-derived drugs returned over $8 billion last year should not go unnoticed.

The Cambridge group, led by Peter Leadlay and Jim Staunton, has focused much of their efforts on the biosynthesis of erythromycin. Sequencing of the erythromycin aglycone biosynthetic genes by the group [3] and independently by Leonard Katz *et al* [4], revealed 3 large open reading frames, each coding for a giant (>3,000 amino acids) multifunctional protein designated deoxyerythronolide B synthase (DEBS) 1, 2 & 3.

6. Modular Polyketide Synthases

Detailed sequence comparisons revealed that the DEBS proteins comprise of some 28 catalytic centres or domains with considerable homology to the enzymes responsible for fatty acid biosynthesis. Importantly, the domains are arranged in 6 modules (2 per DEBS protein), each of which is able to carry out a carbon-carbon bond forming condensation step using 3 essential catalytic activities, ketoacyl synthase (KS), acyl transferase (AT) and acyl carrier protein (ACP), followed by appropriate keto group modification with keto reductase (KR), dehydratase (DH), enoyl reductase (ER) options [5].

The AT domain at the N-terminus of DEBS 1 initiates the carbon bond forming process by transferring a propionyl-CoA primer unit, via the pantetheinyl residue of the first ACP domain, to the KS of module 1 (Fig 1). The Cambridge group [6] also established that (2*S*)-methylmalonyl-CoA is the substrate for the carbon-chain elongation process, this being loaded onto AT_1 and then onto the thiol terminus of ACP_1 where KS_1 then catalyses the first condensation reaction by decarboxylative acylation of the methyl malonyl residue by the propionate starter unit. The resulting ACP thioester is reduced by KR_1 to give a 2-methyl-3-hydroxypentanoyl-ACP diketide intermediate which is transferred over to KS_2 of module 2. The chain elongation process is repeated 5 more times, with varying degrees of keto reduction, until the thioesterase (TE) domain at the end of DEBS 3 catalyses the lactonisation of the polyketide chain and the first enzyme-free intermediate, 6-deoxyerythronolide B (6DEB), is released.

Proof that the individual domains operate in the order shown (Fig 2) came from the Katz group [4] who carried out an 813-bp in-frame deletion in the KR domain of DEBS module 5; the resultant mutant producing 5,6-dideoxy-3a-mycarosyl-5-oxoerythronolide B, plus minor amounts of the corresponding aglycone. The same group [7] also disrupted the enoyl reductase (ER) domain of module 4; the resultant mutant strain producing $\Delta^{6,7}$-anhydroerythromycin C as predicted. The formation of these compounds not only gave direct evidence for the modular nature of the polyketide synthase (PKS) enzymes, but also demonstrated the relaxed substrate specificity of the downstream enzyme domains that were capable of processing unnatural intermediates.

7. Domain/Module Swapping to Produce Novel Compounds

The possibility of repositioning/reassembling the pks genes and, therefore, the coded domains, offers even greater scope for producing new metabolites. The first demonstration of this approach came with the relocation of the TE domain from the end of DEBS 3 the end of DEBS 1 [8].

A mutant of *Saccharopolyspora erythraea* containing the engineered protein (called DEBS 1-TE) gave the anticipated triketide lactone and no erythromycins (Fig 3). In further examples of the relocation strategy by the Katz, Cane and Khosla groups [9], TE has been relocated to the terminus of modules 5 and 3 and expressed in *Streptomyces coelicolor*. Products consistent with truncation of chain elongation at the anticipated stages were obtained in both cases (Fig 4). The truncation at module 5 led to the release of the hexaketide, 12-membered ring macrolide. The truncation after module 3 resulted in the formation of 2 tetraketide products (the pyranone, possibly from the keto acid by decarboxylation after release from enzyme).

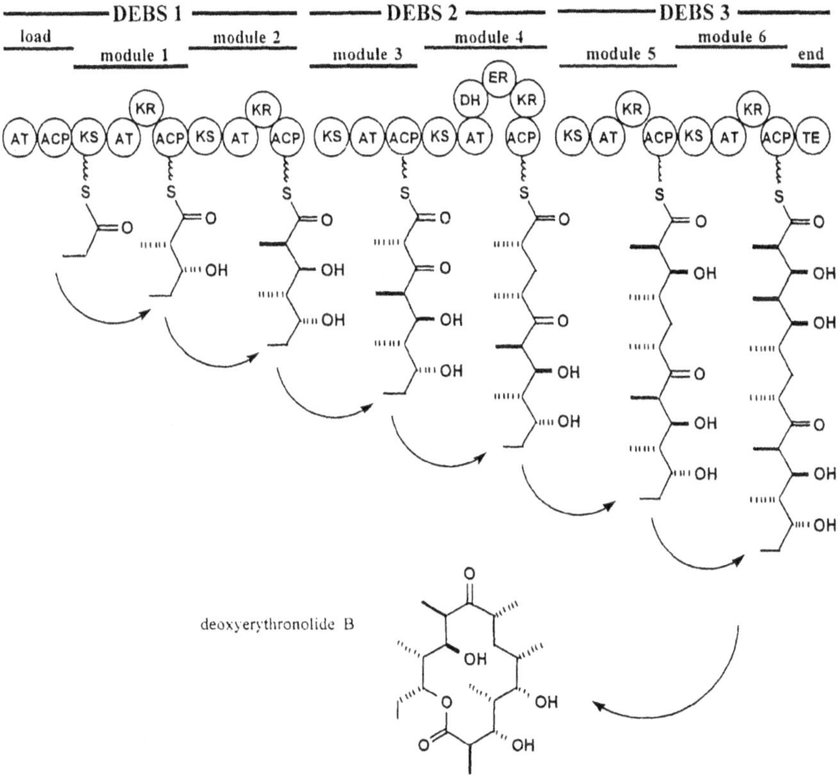

Figure 1. The Erythromycin Polyketide Synthase

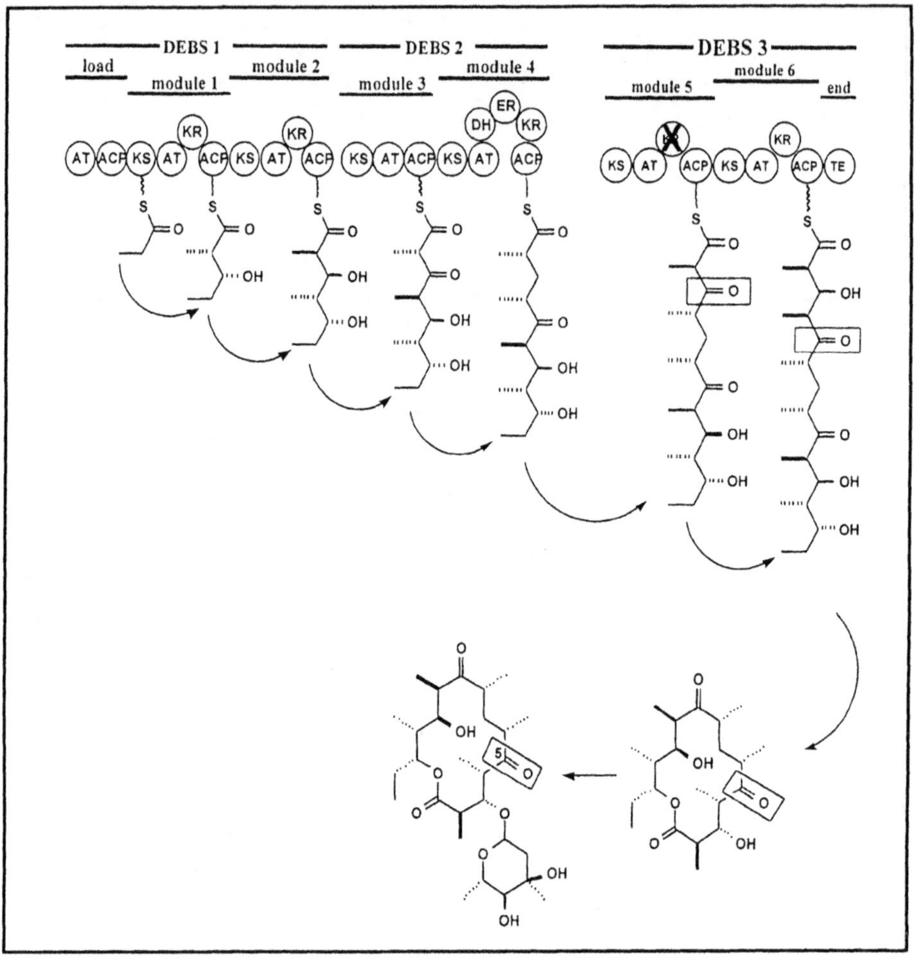
DEBS 1
DEBS 2
DEBS 3
load
module 1
module 2
module 3
module 4
module 5
module 6
end
AT
ACP
KS
KR
DH
ER
TE
S
O
OH

Figure 2.

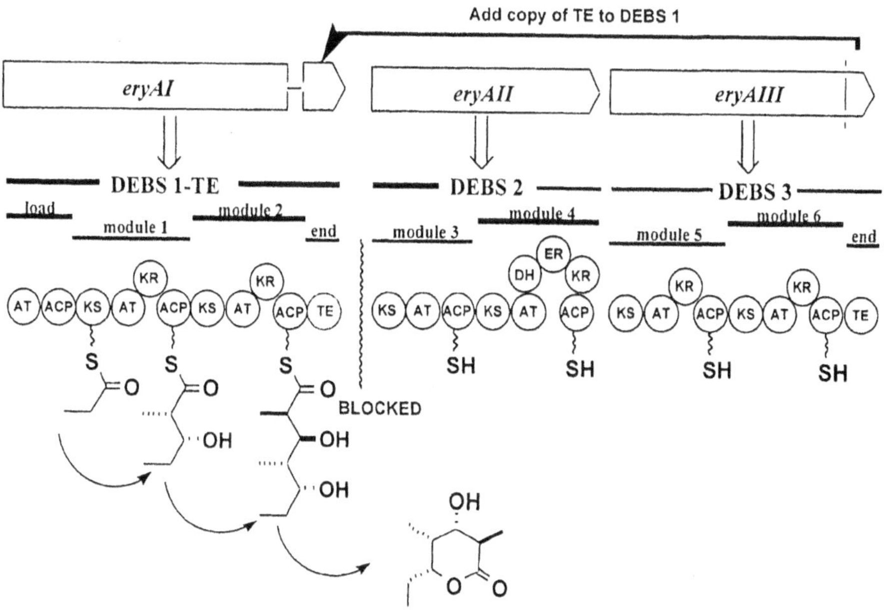

Figure 3. Repositioning thioesterase domain

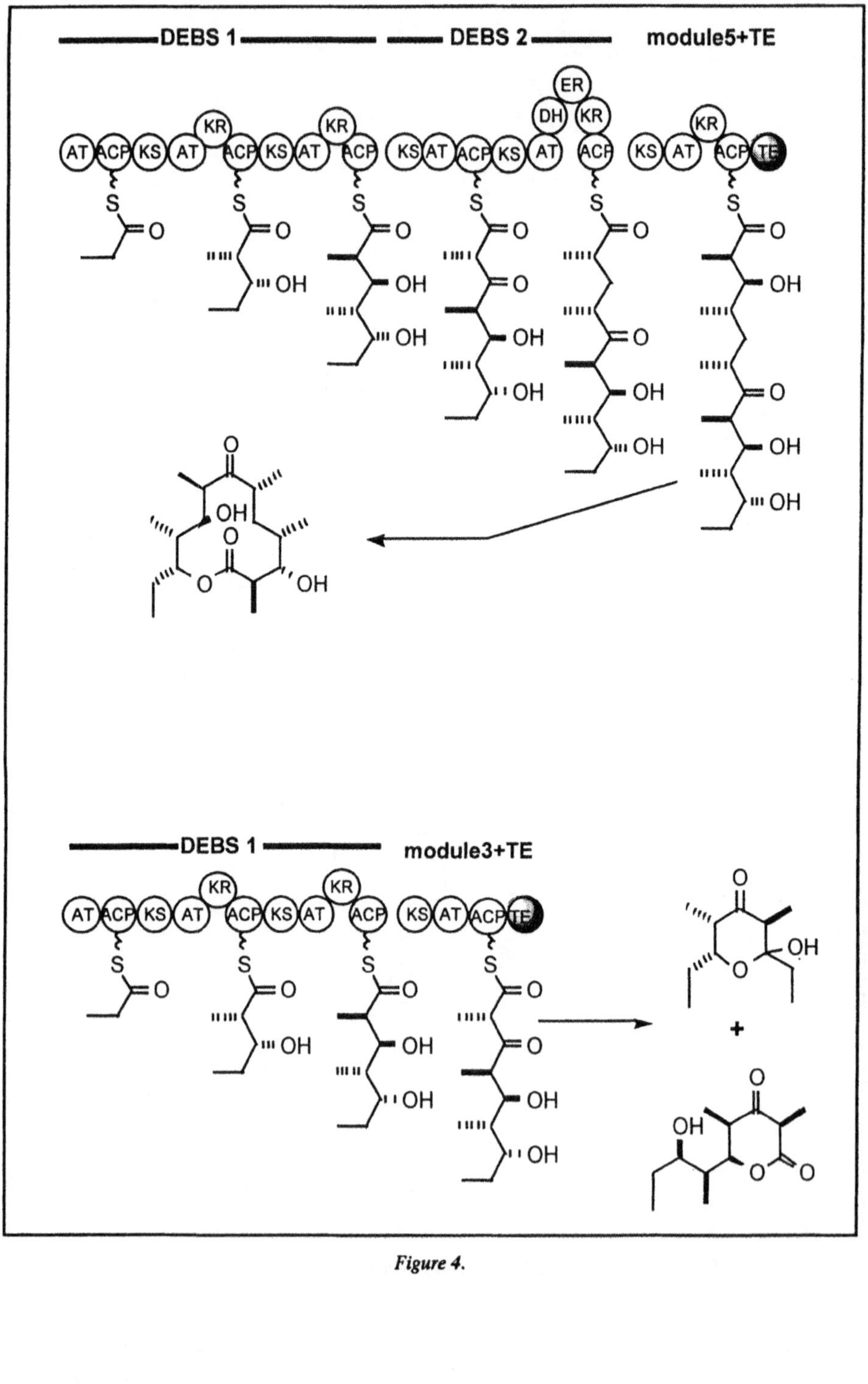

Figure 4.

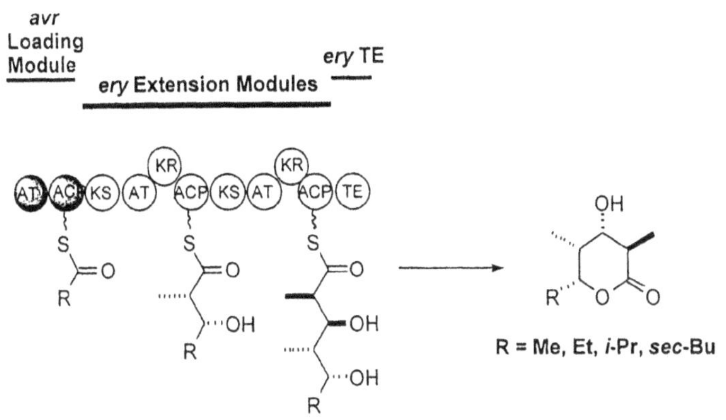

Figure 5. Hybrid triketides by module swapping

The range of new metabolites would be expanded much further if domains could be transferred between *different* PKS clusters ("mix and match" strategy). In an experiment to realise this goal, the Cambridge group [10] hybridised erythromycin and rapamycin PKS systems to produce products derived from the 2 parent polyketides. The experiment involved replacing the erythromycin AT of module 1 by an AT derived from the rapamycin PKS. In its normal function in rapamycin biosynthesis the transplanted AT specifies a malonate unit as chain extender rather then a methylmalonate. The engineered mutant strain produced triketide lactones lacking a methyl group at C-4 (NB: expression in *S. coelicolor* resulted in the lactone analogue with an acetate starter group - presumably due to the varying sizes of intracellular pools of propionyl-CoA and acetyl-CoA in the host strain). A more ambitious heterologous hybrid has also been described by the Cambridge group [11] which involved transplanting a *complete module* rather than a single domain. The loading module (AT + ACP) of the avermectin PKS was transplanted into DEBS 1-TE to produce 4 lactones; 2 were novel hybrid compounds (R= ^{i}Pr and ^{sec}Bu) having starter acyl groups characteristic of the avermectins (Fig 5). Transplanting the avermectin loading module into the complete DEBS system gave novel, 13-substituted (^{i}Pr, ^{sec}Bu) erythromycin A and B analogues [12]! These exciting results demonstrate that domains or modules drawn from different pathways can cooperate to give functional hybrid synthases capable of producing novel metabolites.

Recent work published by the Khosla group [13] used a DEBS mutant expressed in *S. coelicolor*, but inactivated at KS_1. The genetic block was deployed to prevent unwanted competition with propionyl-CoA derived *in situ* by means of enzyme-catalysed decarboxy-

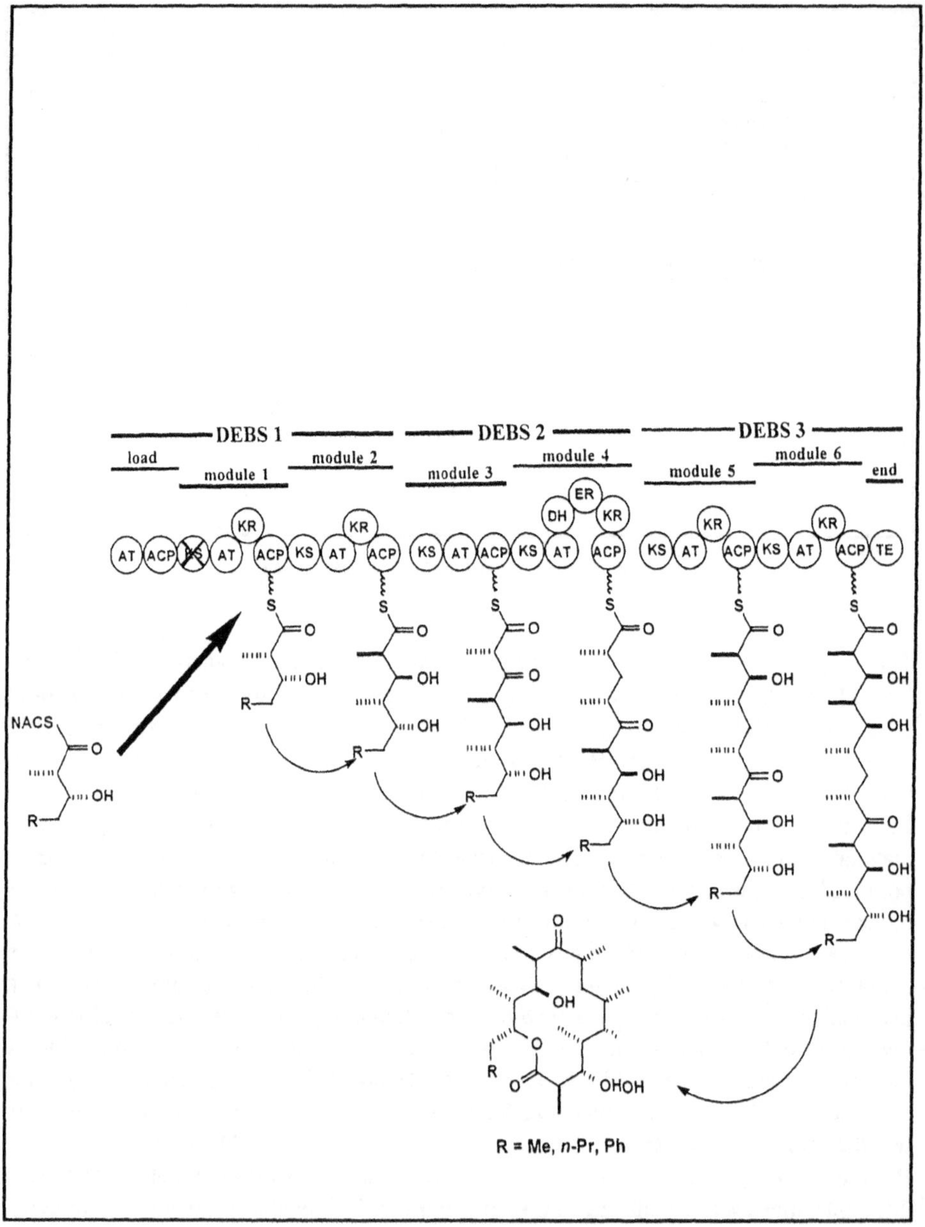

Figure 6. Mutasynthesis of erythromycin analouges

lation of methylmalonyl-CoA. Diketides supplied as N-acetylcysteamine thioesters were added to growing cultures of the mutant and the corresponding 6DEB analogues were isolated (Fig 6). Even a phenyl group was tolerated at the chain terminus to yield the corresponding macrolide.

The successful generation of functional hybrid polyketide synthases and their interception with new, advanced intermediates opens up exciting possibilities for the creation of a wide variety of unnatural natural products with drug-like qualities. The future role of natural products in drug discovery is changing, but a future there most certainly is.

References

1. Patel, D.V. and Gordon, E.M. (1996) Applications of small-molecule combinatorial chemistry to drug discovery, *Drug Discovery Today* **1**, (4), 134-144.
2. Hook, D. (1996) Ultra high throughput screening - a journey into Nanoland with Gulliver and Alice, *Drug Discovery Today* **1**, (7), 267-268.
3. Cortés, J., Haydock, S.F., Roberts, G.A., Brevitt, D.J. and Leadlay, P.F. (1990) An unusually large multifunctional polypeptide in the erythromycin-producing polyketide synthase of *Saccharopolyspora erythraea*, *Nature* **348**, 176-178.
4. Donadio, S., Staver, M.J., McAlpine, J.B., Swanson, S.J. and Katz, L. (1991) Modular organisation of genes required for complex polyketide biosynthesis, *Science* **252**, 675-679.
5. Staunton, J. (1991) The extraordinary enzymes involved in erythromycin biosynthesis, *Angew. Chem. Int. Ed. Engl.* **30**, (10), 1302-1306.
6. Marsden, A.F.A., Caffrey, P., Aparicio, J.F., Loughran, M.S., Staunton, J. and Leadlay, P.F. (1994) Stereospecific acyl transfers on the erythromycin-producing polyketide synthase, *Science* **263**, 378-380.
7. Donadio, S., McAlpine, J.B., Sheldon, P.J., Jackson, M. and Katz, L. (1993) An erythromycin analog produced by reprogramming of polyketide synthesis, *Proc. Natl. Acad. Sci. USA* **90**, 7119-7123.
8. Brown, M.J.B., Cortes, J., Cutter, A.L., Leadlay, P.F. and Staunton, J. (1995) A mutant generated by expression of an engineered DEBS 1 protein from the erythromycin-producing polyketide synthase (PKS) in *Streptomyces coelicolor* produces the triketide as a lactone, but the major product is the nor-analogue derived from acetate as a starter, *J. Chem. Soc., Chem. Commun.*, 1517-1518.
9. Kao, C.M., Luo, G., Katz, L., Cane, D.E. and Khosla, C. (1995) Manipulation of macrolide ring size by directed mutagenesis of a modular polyketide synthase, *J. Amer. Chem. Soc.*, **117**, 9105-9106.
10. Oliynyk, M., Brown, M.J.B., Cortés, J., Staunton, J. and Leadlay, P.F. (1996) A hybrid modular polyketide synthase obtained by domain swapping, *Chem. Biol.*, **3**, 833-839.
11. Leadlay, P.F., Staunton, J., Marsden, A.F.A., Wilkinson, B., Dunster, N.J., Cortés, J., Oliynyk, M., Hanefield, U. and Brown, M.J.B. (1997) in "Industrial Micro-Organisms: Basic and Applied Molecular Genetics", Batltz, R.H., Hegeman, G.D. and Skatrud, P.L., Eds. American Society for Microbiology, Washington DC, *in press*.
12. Leadlay, P.F. and Staunton, J. *personal communication*.
13. Jacobsen, J.R., Hutchinson, C.R., Cane, D.E. and Khosla, C. (1997) Precursor-directed biosynthesis of erythromycin analogs by an engineered polyketide synthase, *Science* **277**, 367-369.

INDEX

Proceedings of the Phytochemical Society of Europe

42. N.J. Kruger, S.A. Hill and R.G. Ratcliffe (eds.): *Regulation of Primary Metabolic Pathways in Plants*. 1999 ISBN 0-7923-5494-X
43. L. Bohlin and J.G. Bruhn (eds.): *Bioassay Methods in Natural Product Research and Drug Development*. 1999 ISBN 0-7923-5480-X

KLUWER ACADEMIC PUBLISHERS – DORDRECHT / BOSTON / LONDON

GPSR Compliance

The European Union's (EU) General Product Safety Regulation (GPSR) is a set of rules that requires consumer products to be safe and our obligations to ensure this.

If you have any concerns about our products, you can contact us on ProductSafety@springernature.com

In case Publisher is established outside the EU, the EU authorized representative is:

Springer Nature Customer Service Center GmbH
Europaplatz 3
69115 Heidelberg, Germany

Batch number: 10809441

Printed by Printforce, the Netherlands